AF844574
10 185 066
...hnic Library
UNIVERSITY OF NORTHUMBRIA
AT NEWCASTLE LIBRARY
WITHDRAWN

92

METAL STRIP FROM POWDER

The modern rolling mill is a highly productive and relatively efficient method of reducing metal ingot to strip: it has a number of disadvantages which stem from the high cost of rolling mill equipment and the high minimum production levels necessary for economic operation.

Numerous suggestions have been made down the years, and techniques developed, in attempts to devise methods of strip production which would eliminate much of the capital equipment required in the conventional rolling mill. Of these, the most promising approach to date has proved to be that of powder metallurgy.

In this monograph a number of methods of making metal strip alternative to conventional rolling are described. The problems, potential advantages and the scientific basis of making metal strip by forming it from powder, sintering and finishing are discussed and the process economics and advantages of the powder route are considered.

A number of individual processes that are in use, coming into use, or under development are described. These range from the concept of directly rolling fine dry powder, through forming a self-supporting uncompacted metal/plastics film for further processing, to the technique of hot rolling coarsely atomised metal which is allowed to fall, or is guided into a roll compaction mill while at high temperature.

The possibility of using the powder route for manufacturing metal foils at economic cost, and for the production of a whole range of new metallurgical products and composite materials suggests that powder strip technology could greatly extend the practical applications of thin metal strip.

Other titles in this series

ME/1 TRIBOLOGY
E D Hondros

ME/2 ELECTRON BEAM WELDING
M J Fletcher

ME/3 VACUUM BRAZING
M J Fletcher

ME/4 STRESS CORROSION FAILURE
P Greenfield

ME/5 GLOW DISCHARGE MATERIAL PROCESSING
R A Dugdale

ME/6 DIAMOND GRINDING
J Burls

ME/7 ENGINEERING APPLICATIONS OF BERYLLIUM
P Greenfield

ME/8 WHISKERS
C C Evans

ME/9 CREEP OF METALS AT HIGH TEMPERATURES
P Greenfield

ME/10 CAVITATION
I S Pearsall

ME/11 MAGNESIUM
P Greenfield

ME/12 ZIRCONIUM IN NUCLEAR TECHNOLOGY
P Greenfield

ME/13 FRICTION WELDING
M J Fletcher

M&B Technical Library

TL/ME/1 HYDROSTATIC EXTRUSION
J M Alexander and B Lengyel

TL/ME/2 THE HEAT PIPE
D Chisholm

TL/ME/3 PLASMA SPRAYING
R F Smart and J A Catherall

TL/ME/4 FATIGUE
J T Barnby

All published by Mills & Boon Limited

M & B Monograph ME/14

General Editor: J Gordon Cook, PhD, FRIC

Metal Strip from Powder

G M Sturgeon, C Eng, M I Mech E
R L S Taylor
British Steel Corporation

Mills & Boon Limited
London

First published in Great Britain 1972
by Mills & Boon Limited, 17–19 Foley Street,
London, W1A 1DR

© Mills & Boon Limited 1972

ISBN 0.263.05113.7

All rights reserved. No part of this publication may be reproduced, stored in a retrieval system, or transmitted in any form or by any means, electronic, mechanical, photocopying, recording or otherwise, without the prior permission of Mills & Boon Limited

NEWCASTLE UPON TYNE
POLYTECHNIC
LIBRARY
ACCESSION No. 102080
CLASS No. 621.762

Made and printed in Great Britain by
Butler & Tanner Limited, Frome and London

CONTENTS

FOREWORD

In writing a short introductory monograph on an active and rapidly changing area of science and technology it is difficult for the authors to be certain that they have presented a fair and reasonably balanced view of the state of the art. No one can know everything taking place in his field, the authors wish to thank the many scientists, technologists and engineers who have generously helped and commented on the book during the course of preparation. Any undue emphasis or error is entirely the responsibility of the authors.

We would like to record our thanks to Dr R. S. Barnes, Director of Research and Development, British Steel Corporation, for permission to publish details of work done by the Corporation on the manufacture of strip from powder. In addition, we would like to thank the Iron and Steel Division of the Department of Industry, Trade and Commerce, Government of Canada and colleagues in the British Steel Corporation who kindly assisted with the illustrations, in particular, Mr D. L. Smith and his staff at the British Steel Corporation Corporate Laboratories, Battersea, London.

1. Introduction

METALS AND MAN

Our civilisation is founded upon metals and their use. Prior to the discovery of metals, man had developed techniques for working other materials such as flint, bone and wood, and had acquired a not inconsiderable culture. Our knowledge of metals, their properties and uses, is likely to remain the central significant fact of our civilisation.

Metals play a vital role in almost every aspect of modern life. We use metals for making planes and ships, cars and trains, buildings and bridges. Our manufacturing industries depend on metal machinery; we grow our food with the aid of metal tools. It is not surprising, therefore, that the techniques of manufacturing and forming metals have a profound effect upon the economic life of the world. The continuous refinement of existing techniques and the development and application of new methods of making or forming metals are essential to the future progress and wellbeing of mankind.

In this book, we have considered one aspect of metalworking technology: the manufacture of thin metal strip from metal powder. We have concentrated, in particular, upon the production of steel strip, because steel is the most important metal in general use, and one which at the same time poses severe problems for the scientist, engineer and technologist seeking to develop bulk tonnage powder metallurgical processes for the manufacture of strip.

NEW ROUTES TO METAL STRIP

Thin metal strip is one of the major raw materials of our time, and is thus of immense significance economically. Mild steel strip, for example, finds applications in a wide range of products, from motor cars to "tin" cans;

stainless steel and non-ferrous metal strip are used for an equally wide range of products.

Fig. 1. A steel ingot being drawn from the soaking pits where the material has been brought to the desired uniform temperature for rolling to be commenced. (Port Talbot Works, Strip Mills Division, BSC.)

In general, strip is made from these metals by casting a massive ingot (Fig. 1) and then subjecting it to a series of reducing processes by rolling (Fig. 2). Although little metal is wasted, the route from a rough ingot to the final gleaming roll of coiled strip is so capitally intensive that costs are high.

Throughout the present century, less capitally intensive ways have been sought of making steel and other metal strip without losing process efficiency or adversely affecting the properties of the finished material. In recent years, techniques which have been investigated include the following:

Fig. 2. The final stage of conventional strip rolling. The six-stand finishing mill at the BSC works, Llanwern.

(1) Float strip;
(2) Continuous casting;
(3) Spray rolling;
(4) Compaction of metal powder.

(1) *Float Strip*

The float strip process, illustrated schematically in Figure 3, is a development of the technique used so successfully by Pilkington Brothers Ltd. for the manufacture of glass. Liquid metal is allowed to float upon another molten metal or material of higher density and lower melting point. Ideally, the floating metal forms a uniform layer which can then be cooled and removed from the bath,

Fig. 3. Direct strip production 'float' process.

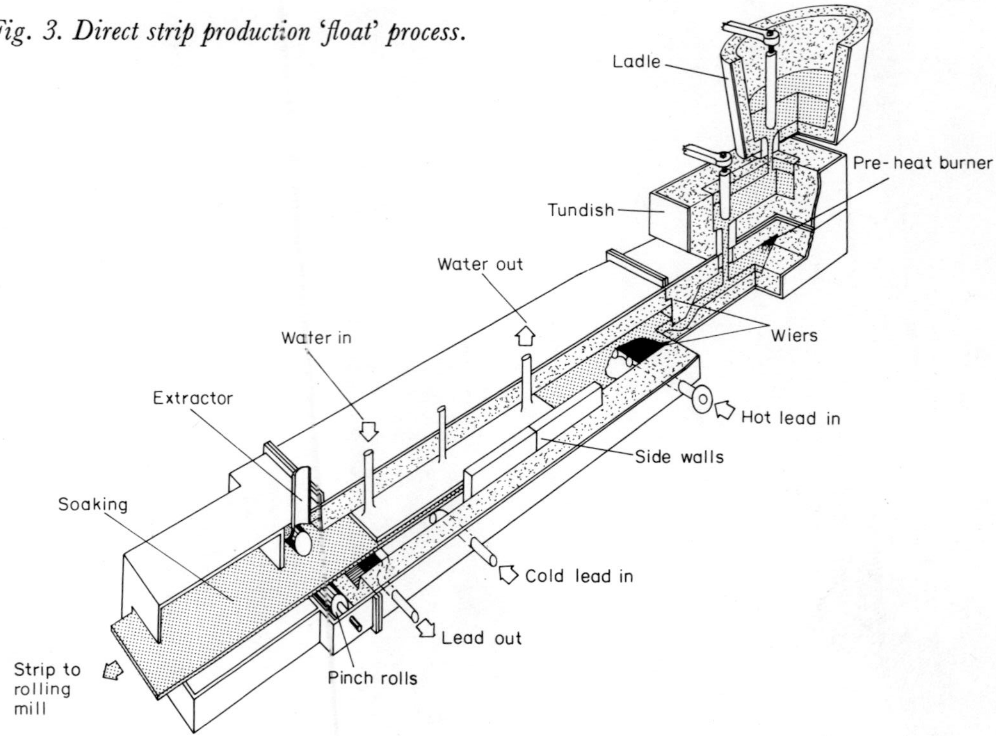

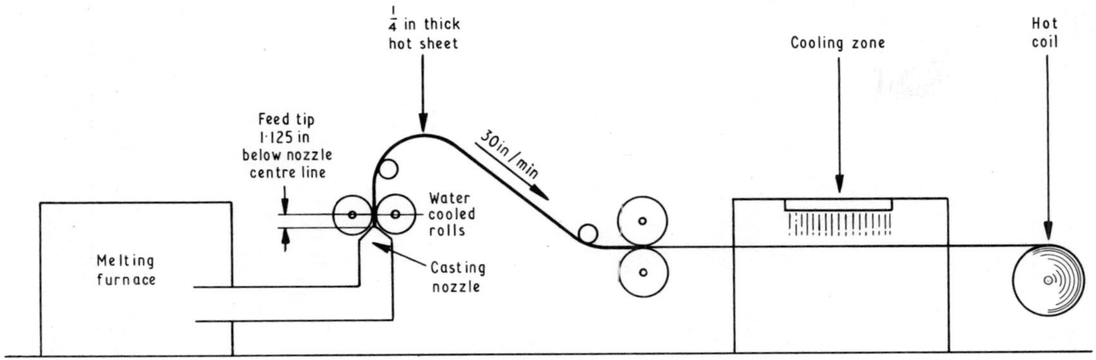

Fig. 4. Continuous strip casting. The Hunter process used for non-ferrous alloys.

which is then available for the production of a further length of strip. The process may be operated continuously by pouring in liquid metal at one end of the flotation tank, progressively cooling the floating strip along the length of the tank, and finally withdrawing it at the other end; hot strip is thus produced continuously, ready for further rolling or processing.

Unfortunately, this technique has met with little practical success as a means of producing metal strip. There are many problems; for example, the floating metal tends to form an alloy with the metal on which it floats, or to absorb some of the material from it; also, it is necessary to cool the floating strip rapidly to prevent vaporisation of the flotation medium in the bath, and thus avoid chemical contamination of the strip through the formation of complex alloys and oxides.

(2) *Continuously Cast Strip*
The continuous casting of strip into a specially cooled mould, from which material is withdrawn either in the form of plates or continuously as strip, has already come into commercial operation, at least for non-ferrous materials. Considerable advances have been made in designing semi-mouldless machines for continuously carrying out the strip casting of aluminium and its alloys, and

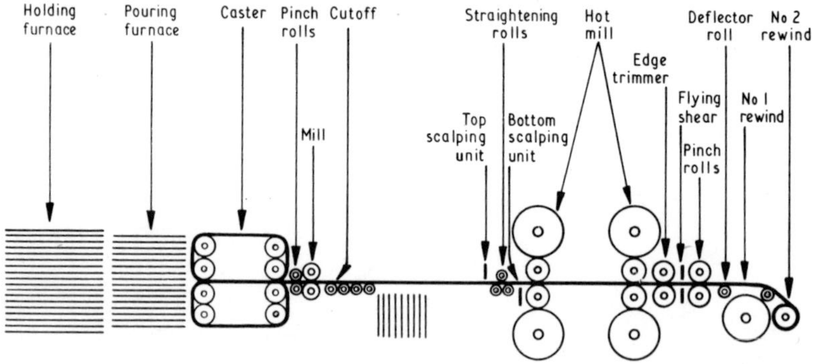

Fig. 5. Continuous strip casting. The Hazlett process for aluminium has been investigated and considered as a possible technique for steel.

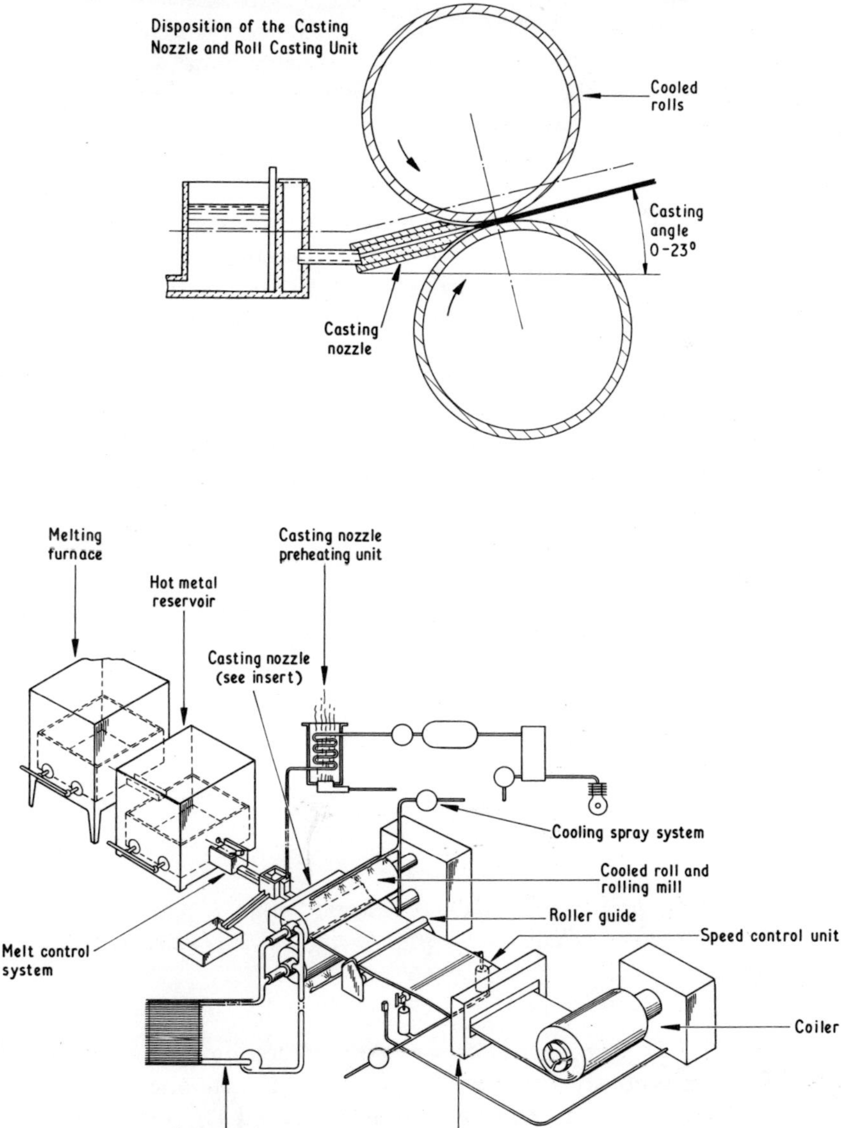

Fig. 6. The Harvey continuous strip casting system is claimed to be suitable for a wide range of metals, although it has been primarily developed for aluminium.

a number of variants of this process exist, such as the Hunter,[1] the Hazlett[1] and the Harvey[2] machines (Figs. 4, 5 & 6). Research and development work is continuing on the continuous casting process, and strenuous efforts have been made to extend the techniques to the steel industry.

In the case of steel it is necessary to operate at much higher temperatures, and the surface tension is lower; these facts pose real problems for the development engineers, but it seems probable that a continuous casting technique such as squeeze casting[3] shown in Figure 7 will eventually become commercially viable for plate and light plate gauges of steel. Much of this material would then be rolled conventionally down to thinner gauges, eliminating at least part of the present-day processing cycle, with consequent economic advantage.

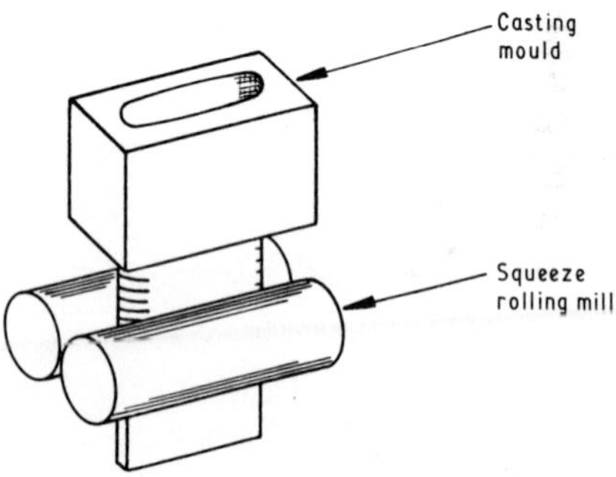

Fig. 7. Squeeze cast thin slab. A flattened "oval" is formed into slab by squeeze rolling shortly after leaving the mould and before cooling.

Electrostatic Continuous Casting An interesting technique for continuously casting steel strip, which in one sense foreshadows spray rolling, was investigated by the Steel Company of Wales[4] (now part of the BSC).

The technique involves imparting a high electrostatic potential to a bath of molten metal. The metal is then caused to flow through a nozzle of suitable design into a confined space, which contains an inert atmosphere such as argon. The design of the nozzle and the high electrical potential (80,000 volts up to 200,000 volts) causes the stream to atomise into a fine spray. This spray is collected on a cooled moving surface, where the droplets coalesce to form a steel strip. This is subsequently peeled from the receiving substrate, and given one or more conventional roll passes prior to coiling. In the schematic diagram (Fig. 8) the mode of collection is to deposit and form the

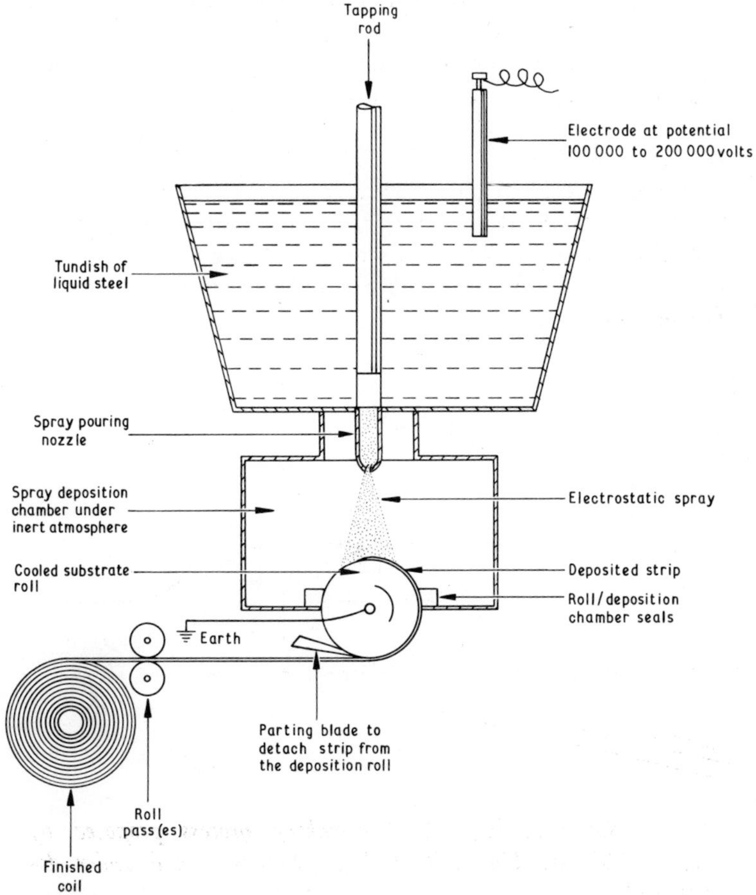

Fig. 8. Electrostatic continuous strip casting process. Developed by the Steel Company of Wales (now part of the Strip Mills Division, BSC).

strip on the surface of a cooled roll. With this type of layout, care would be necessary to prevent excessive loss of the inert gas atmosphere, but the advantages gained in plant compactness would be considerable. However, the main problems the process will have to overcome to gain acceptance are psychological rather than technical, as there is little doubt that the high electrostatic potentials required are a new factor in tonnage steel production and people have yet to be convinced that acceptable safety standards can be maintained. For this reason, it seems probable that a rather more conventional approach, such as those described earlier, may be adopted eventually for strip production by continuous casting.

(3) *Spray Rolling*

The spray rolling[5] of metal (Fig. 9(a), (b)) has been pioneered by A. R. E. Singer, Professor of Metallurgy at the University College of Swansea. It is currently under investigation with the support of the National Research Development Corporation.

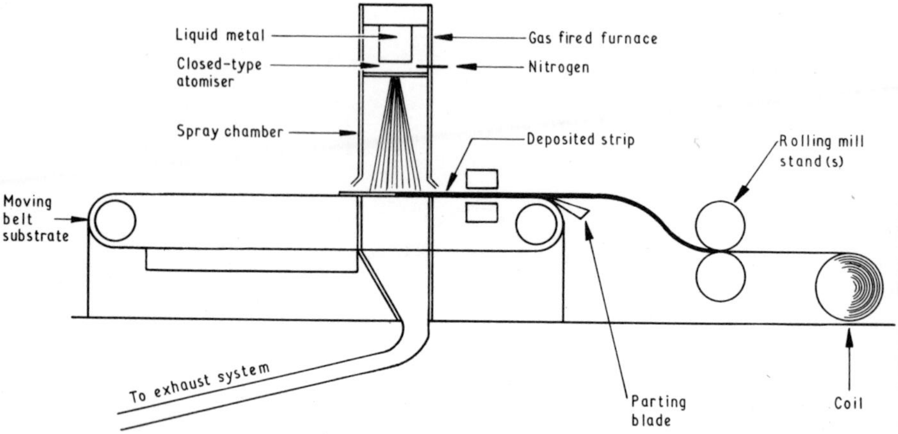

Fig. 9. Spray rolling. A strip-making process proposed by Professor Singer, University College, Swansea, and under development.
(above) Schematic of spray deposition using a flat substrate and a closed type atomiser.
(right) Schematic of spray deposition using a roll substrate and open type atomiser.

In spray rolling liquid metal is atomised in a specially constructed chamber with an inert gas such as nitrogen. The spray of atomised particles is directed onto a substrate; this has been prepared in such a way that the first layer of atomised particles will adhere to it, to form a smooth coating, but the thickened strip may be peeled from the substrate while still at high temperature and hot rolled to give fully dense strip. All these stages are carried out under an inert gas atmosphere, but finishing can be done by normal rolling procedures and the metal strip has properties similar to those made by conventional means.

Theoretically, the process is attractive, but in the case of steel strip manufacture (for which the process was originally conceived) there are many practical problems of atomisation, thermodynamics, engineering and plant main-

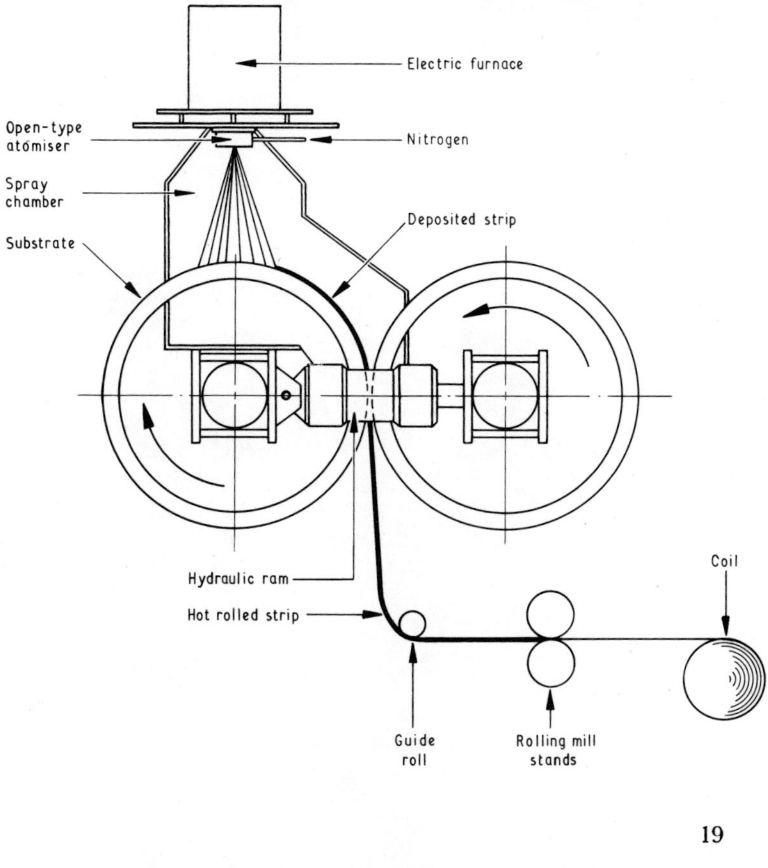

tenance. Time will be needed to enable these problems to be solved in the case of steel strip. The technique could be successfully applied rather earlier to the manufacture of light non-ferrous alloy materials. The most crucial problem is that of getting the initial deposit to form a sufficiently uniform film to yield satisfactory finished strip.

(4) *Strip from Powder*

The technique of producing strip from powder seeks to reverse the processing cycle of the conventional rolling process in that metal strip is made by continuously compacting, heat treating and working fine powders in such a way that the particles coalesce into solid metal. This technique of producing strip from metal powder has appeared for many years to offer the prospect of lower capital and operating costs, but all the earliest attempts

Fig. 10. Processing strip from powder at the Hoyle Street (Sheffield) laboratory of the BSC Corporate Laboratories.

to put it into practice ran into severe technical problems; the manufacturing processes developed were unable to compete in efficiency with those used in the best large rolling mills. Currently, however, the position has changed; powder rolling has been accepted as a practical commercial process for the small scale production of premium grade materials (Fig. 10) and has been demonstrated to be commercially viable for much larger tonnage production. Consequently, the first tonnage commercial application of the technique is imminent.

2. Why Strip from Powder?

Before discussing some of the techniques that have been devised for turning metal powders into strip, it is worth asking why strip from powder should be competitive with the modern rolling mill.

The development of conventional rolling mills of all types has been surprisingly rapid. Large integrated plants, automated and computer-controlled, are now used for rolling steel and other materials in a wide range of gauges, and there has been a continuing demand to cut costs and improve the properties of the steel and other metals produced. Unfortunately, as plants become more sophisticated and increase in size, it becomes increasingly difficult to match their output to market demand. An appreciable increase in demand for the materials they produce means a formidable investment programme to provide plant for ingot casting, rolling, annealing and other associated processes. Because of this, decisions to increase capacity have tended to be postponed until the market has expanded to a point at which it can take up the major proportion of the added output, as the increments by which capacity is increased in the case of modern strip mills are of the order of 3 million tons a year. Because of this dilemma the metallurgical industries have tended to see-saw between an inadequate and an over-adequate production capacity.

There is an obvious attraction in the powder route to strip production, which can eliminate the casting of ingots, the hot rolling of ingots and slab, and much of the cold rolling, together with annealing and other ancillary processes (Fig. 11). The lower capital cost of powder strip plant would mean that they could provide a more effective way of keeping supply in step with demand.

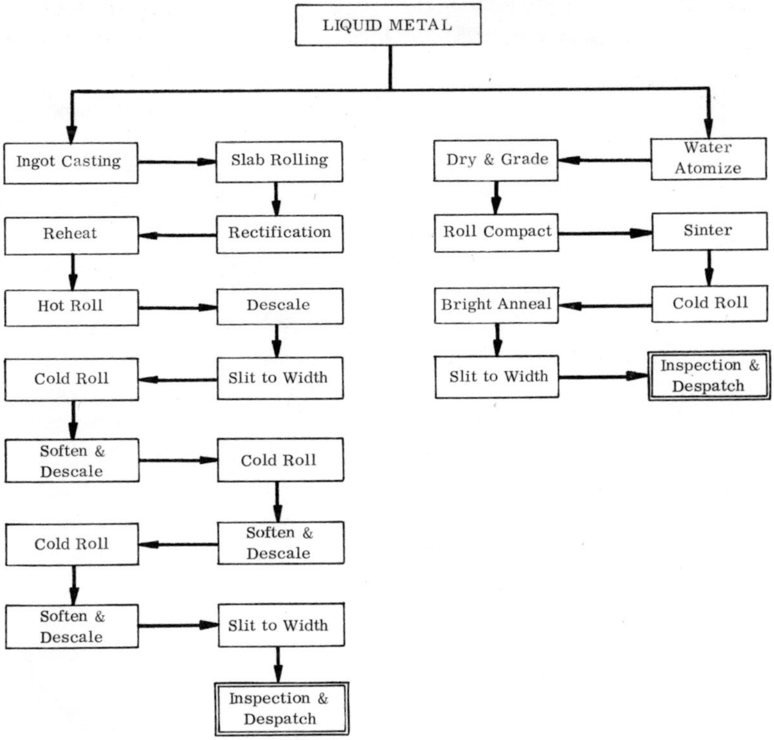

Fig. 11. Comparison of conventional and powder production routes for stainless steel strip.

THE BEGINNINGS OF POWDER ROLLING

The production of metal strip from powder probably began in 1843, when Henry Bessemer inadvertently made brass strip while feeding fine turned brass particles into a rolling mill, with the intention of making brass flake for use as a pigment in "gold" paint. Sir Henry Bessemer's[6] description of what happened is as follows:

"This stream of filaments* was conducted between the rolls matted and felted together in inextricable confusion, and in this state, they had a strong tendency to unite and

* The brass turnings.

so weld themselves together under pressure as to issue from the rolls with a smooth continuous surface, resembling an ordinary sheet of solid brass. This would soon become too compact to separate and break up again, . . ."

Henry Bessemer's main aim was to manufacture brass flake, and he did not appear to appreciate the commercial possibilities of a process for making strip from powder. It was not until 1902, almost sixty years later, that a patent embodying the idea was filed by Siemens & Halske[(7)] in Germany. In 1950, G. Naeser and F. Zirm[(8)] described a process in which iron powder was compacted between a pair of rolls to form what is termed "green strip"; that is, strip composed of particles of powder sufficiently bonded together to constitute a continuous but easily friable band. This was then sintered, and the resulting material processed conventionally by cold rolling and annealing.

DIRECT ROLL COMPACTION

Various methods of producing strip from powder have been described for a number of metals. One method which has been widely applied to quantity production is "direct roll compaction". The mill used in this process is similar to a conventional rolling mill, but is usually turned on its side, so that the rolling direction is vertically downwards; this avoids the difficulty of feeding loose powder sideways into the roll nip.

Two basic methods of supplying powder to the roll nip used in direct roll compaction can conveniently be described as unsaturated and saturated feed systems. In the unsaturated feed system, an external device is used to meter the powder into the roll nip, so that very little, if any, excess powder is present in the roll nip. In the saturated feed system, on the other hand, the roll nip is flooded with powder.

The main parameters affecting compaction by direct

powder rolling are powder characteristics such as particle shape, size distribution and inter-particle friction, together with the plant parameters—roll diameter, roll surface finish and the area of the roll exposed to the powder to be compacted. In the case of the saturated feed system, these plant parameters are of particular significance, as they have an important role in achieving uniform powder distribution and "green" strip density; they also affect the maximum thickness of strip that can be produced on a particular mill. Generally speaking, the ratio of the compaction mill roll diameter to the maximum strip thickness that can be compacted satisfactorily on a particular mill is of the order of 200 : 1 for ferrous metal powders, although this varies to a greater or lesser extent with the characteristics of the powder. The range of thicknesses that can be compacted, however, can be extended beyond this if lubricants are added to the powder prior to compaction.

Unlike conventional rolling, in which the aim is to reduce one variable, namely strip thickness, two variables, thickness and density, have to be controlled during roll compaction. Strip thickness depends upon the gap between the rolls but any change in the gap will also increase or decrease the density of the strip unless steps are taken to control the amount of powder being compacted. This is usually achieved by controlling the area of the roll surface in contact with the powder, which controls the pumping action of the rolls.

Until recently, there was a further limitation in that rolling could not usually be done faster than 2–10 metres per minute, although a speed of 20 metres per minute with stainless steel powder was reported as early as 1954 by S. Storcheim and co-workers in the U.S. Atomic Energy Commission.(9) Working with copper powder, D. G. Hunt and R. Eborall(10) reached a speed of 23 m/min in 1960, and S. R. Crooks(11) achieved a rolling speed of 30 m/min with iron powder in 1962. A major factor limiting the speed of direct roll compaction is the expulsion of occluded air from the region of the roll nip. Higher speeds or thicker strip are possible by compacting in low viscosity gas such as hydrogen, or in vacuo, or by accelerating the powder particles into the roll nip.

There are thus a number of problems which must be met and solved in direct rolling of metal powder to strip. Despite this, variants of the direct rolling process have already found commercial application for the production of strip from expensive or difficult metals or alloys, and in some cases with softer or more ductile materials where the economics of the situation do not necessitate high rolling speed. But up to the present time, in every case where the operation has been profitable, the finished strip has been a special product able to command a premium price. Indeed, manufacture is not always possible by conventional techniques, as is the case with refractory metals such as tungsten, molybdenum, tantalum and niobium, all of which are now commonly processed into sheet or strip form from powder.

Cobalt Strip: Pfizer Process

The Chemicals Division of Pfizer Ltd.[(12)] introduced ductile and high purity cobalt strips in 1968, as the first of a range of direct powder-rolled products for special applications. Although the rolling speed of their process is known to be low, it has proved a practical proposition for Pfizer to market strip in widths of up to 0·2 m, and of thicknesses ranging from about 0·025 to 1·3 mm at a density claimed to exceed 98% of theoretical value in the finished product. Cobalt strip produced by this method is said to possess excellent ductility and formability, combined with an elongation of some 50%.

Sherritt Gordon Process

Other companies, including Sherritt Gordon Mines Ltd.,[(13, 14)] International Nickel[(15)] and Manganese Bronze[(16)] are engaged in the manufacture or production research of premium grade strip from expensive metal or alloy powders. In fact, Sherritt Gordon Mines Ltd., of Fort Saskatchewan, Alberta, Canada, have been producing pure nickel strip from powder since 1961; it is suitable for the manufacture of coinage by the Canadian and other mints. The Sherritt Gordon process consists of powder blending, followed by the direct roll compaction of the nickel powder through a 64 cm roll diameter mill, to produce green strip approximately 4·25 mm thick. The strip 36 cm wide is produced at a line speed of about 12 m/min.

Sintering of the green compacted strip is done under a hydrogen atmosphere, and the sintered product is treated to remove sulphur. Finally, the strip is hot rolled at 900°C, with a 50% reduction in thickness, and then cold rolled to the required gauge.

Fig. 12. Canadian coinage minted from direct roll compacted nickel powder strip. The actual size of these coins can be gauged from the British 5p coin (left) which is struck from conventionally produced cupro-nickel strip.

Sherritt Gordon sell the material as it stands, or further process it into coin blanks or finished coin (Fig. 12). It is believed that they are currently producing something of the order of ten million pounds of nickel and cobalt strip per annum, and that their investment in powder metallurgical strip production is showing a very favourable return. In recent years, the company has been engaged in the production of special alloys and dispersion-strengthened materials for the NASA space programme, and it is thought at least some of these have been produced by a powder metallurgical route.

Tool Steels and other metals

At least two U.S. steel companies have been exploring the powder route for the production of tool steel. In the case of the softer and more ductile metals, direct powder rolling has been investigated as a production route for aluminium, copper and some precious metals. However, every one of these processes is subject to constraints which would severely limit or, in many cases, prevent their use as the basis of a large tonnage production process for mild or stainless steel. This is because they do not solve many of the problems mentioned earlier; instead, they merely avoid them by working within the restrictions of a relatively unsophisticated technological approach.

HOT POWDER ROLLING

There are two other techniques of roll-compacting powder into strip, both of which have advantages and disadvantages for certain usages. These are the *slurry rolling process*, such as the BISRA* TSP route, and *hot powder rolling*.

Hot powder-rolling techniques make it possible to use cheap gas-atomised[17] spherical powder particles, whereas both direct powder rolling and the slurry rolling process, where cold compaction is generally used, call for powder particles of irregular shape such as those produced by water atomisation[18, 19] or certain chemical reduction routes[20, 21, 22, 23, 24]. This is in order that a good mechanical lock can be achieved between the particles in the gas-compacted green strip, so that it can be satisfactorily handled through the sintering stage of the processing cycle. For bulk tonnage production of powder, however, it appears likely that gas atomisation of a stream of liquid metal could offer some advantages over water atomisation in respect of control and the freedom from oxide contamination. Unfortunately, the droplets produced when a stream of molten metal is gas atomised have sufficient time for surface tension to take effect between the atomisation and final solidification to powder, in spite of the variation of individual droplet size. As a result, gas-atomised powders may be likened to extremely fine lead

* Now BSC Corporate Laboratories.

shot, in that all but very few of the particles will be spherical, and those that are not will in all probability be agglomerations of spherical particles.

Such powders rolled cold will not give a satisfactory mechanical lock in the green strip state, and this holds true even if the rolling operation is done at a very high load in order to flatten and attempt to cold-weld the particles together. As two of the main aims of the powder metallurgical route are to reduce as far as possible the capital costs involved and the amount of work needed to achieve the finished product, it is clearly desirable for both these reasons to do the rolling operation at a load no higher than that needed to achieve the required properties in the finished strip. In fact, the loads needed to produce the degree of compaction necessary to cause cold welding would create mill design problems and result in greatly increased capital cost. This has led to the concept of processing gas-atomised spherical powder particles while they are still hot and malleable, performing atomisation, transportation and compaction of gas-atomised powders in a continuous sequence.

As yet no one has reached the stage from which commercial operation would be the next step. However, processes which closely approach such a continuous sequence have been investigated by the Reynolds Metal Company[(25)] for aluminium and aluminium alloy strip, by the Republic Steel Corporation[(26)] in the United States, and by Schloemann AG in West Germany for steel. It is not yet possible to say if either of these processes will finally prove to be commercially successful, but a number of similar techniques are being examined in the U.K. and in other countries. If any are successful, quite substantial gauges of sheet and plate could be produced economically by a hot powder compaction process.

3. Direct Roll Compaction of Powder

In the basic process for the production of strip by the direct roll compaction of powder, the fine, irregular powder is allowed to fall from a feed-hopper into the roll-nip. It is drawn through the nip partially by friction between the powder and roll surface, that is by a form of pump action, and partially by gravitational force. Subsequent processing involves sintering, further cold reduction, an additional sintering or annealing operation and—if required—one or more temper rolling passes to produce the final finished product.

Although at present exploited for small lots of special qualities only, the following description relates the development of the direct powder rolling process for stainless steel at the corporate laboratories of the British Steel Corporation at Hoyle Street, Sheffield.

THE DIRECT POWDER ROLLING OF STAINLESS STEEL STRIP

The powder-processing cycle (Fig. 11) consists of atomising the liquid steel with high-pressure water, followed by screening and drying the powdered metal. This is then compacted in a rolling mill to a thickness dependent on the final properties and gauge required, sintered in a hydrogen atmosphere, cold rolled to size with a reduction of thickness of 30 to 60%, and finally annealed. Whilst there are a number of alternative routes which may be considered potentially commercially viable for the manufacture of iron or mild steel powders, in the authors' view the most attractive route at present for alloy steel powder production, where the compaction stage is to be carried out cold, is water atomisation.

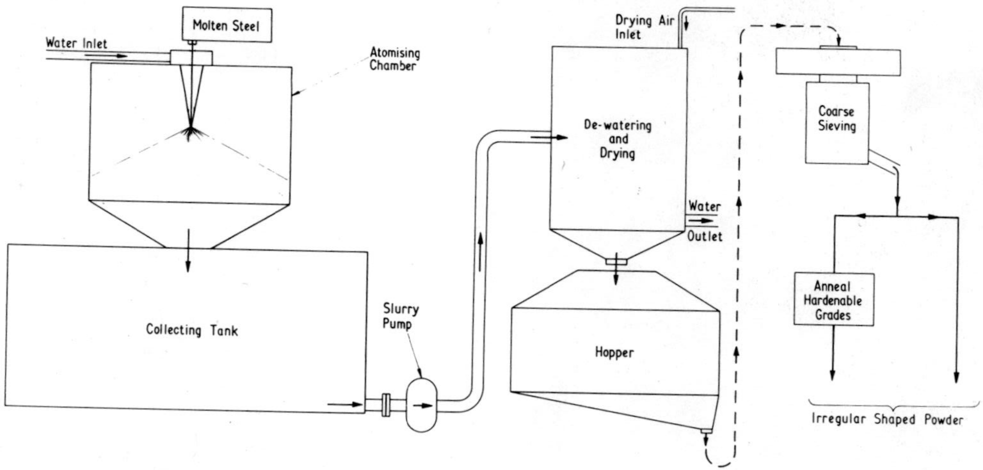

Fig. 13. Atomisation plant.

Fig. 14. BSC stainless steel water-atomisation powder production experimental plant, Sheffield.

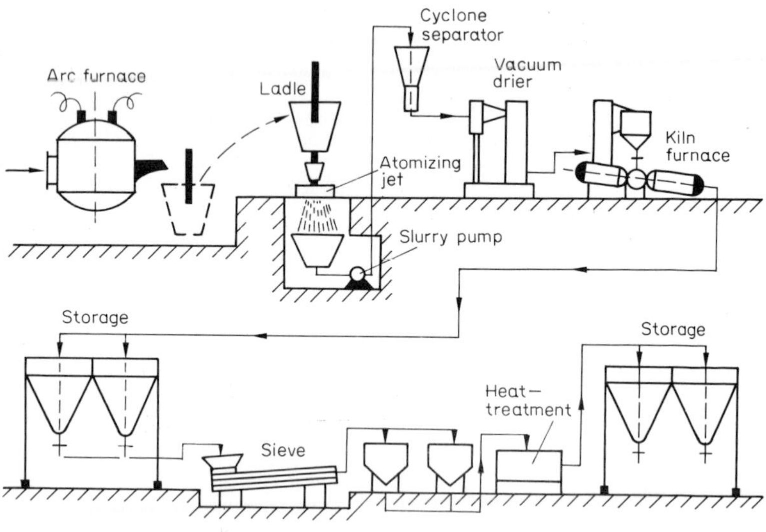

Fig. 15. The Mannesmann atomisation plant for the production of moulding grade powders.

Figure 13 shows a water atomising process suitable for the production of stainless steel powder. It will be seen that the atomisation stage of the process is, in fact, relatively simple and basically cheap. Figure 14 shows the experimental BSC powder production unit at Sheffield. The high price of currently available atomised moulding grades of powder is due to the fact that the main production costs occur during the subsequent blending and grading of the powder to meet customer requirements and specifications (Fig. 15). (17, 19)

The difference in structure and appearance of water- and gas-atomised powders and the various types of powder (17 to 24) produced by other techniques can be seen in the stereoscan electron micrographs in Figure 16. Studies of atomisation have shown that the yield of usable powder depends both upon the atomising conditions and on the average particle size required. For foil manufacture, requiring very fine powders, the yield is obviously lower than for thicker gauges of strip, where coarser powder particles can be used. A typical particle size distribution of a water-atomised stainless steel powder suitable for powder rolling is shown in Figure 17.

The properties required in any given powder for the production of strip are, of course, dependent upon the processing cycle to be adopted. As will be seen later, in hot powder compaction, large spherical particles can be compacted satisfactorily into strip, both for ferrous and non-ferrous metals. For cold compaction, both in the case of direct powder rolling and the slurry technique, an irregularly-shaped particle is required to assist in mechanical locking of the particles during compaction, and to impart the maximum green strength to the compacted strip so that it can be handled satisfactorily through the sintering stage of the processing cycle. Particle size and size distribution are of course relevant to the finished thickness of the strip, and to the amount of work that is necessary to give the final material the properties required. The majority of work in the strip gauge range 0·12 to 3 mm done at the British Steel Corporation's

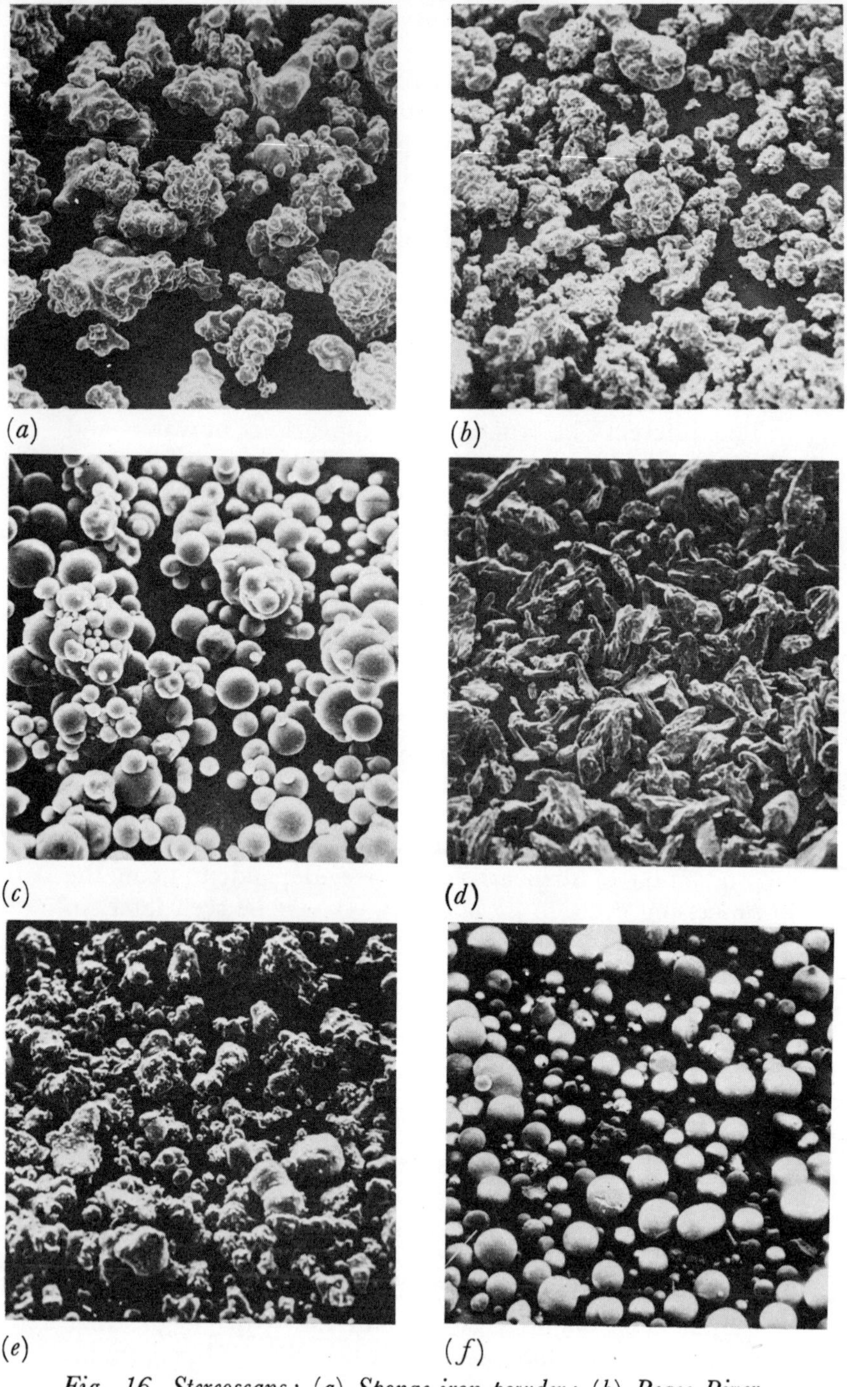

Fig. 16. Stereoscans: (a) Sponge iron powder; (b) Peace River chemically produced iron powder; (c) carbonyl iron powder (mag. 10× greater); (d) electrolytic iron powder; (e) water-atomised mild steel powder; (f) gas-atomised mild steel powder.

Sheffield laboratory, has been undertaken on a 150 micron to dust powder, but there is little doubt that for thicker sizes of strip, the particle size range can be extended to a possible limit of 500 microns, provided that the powder does not segregate during transportation or in the roll nip.

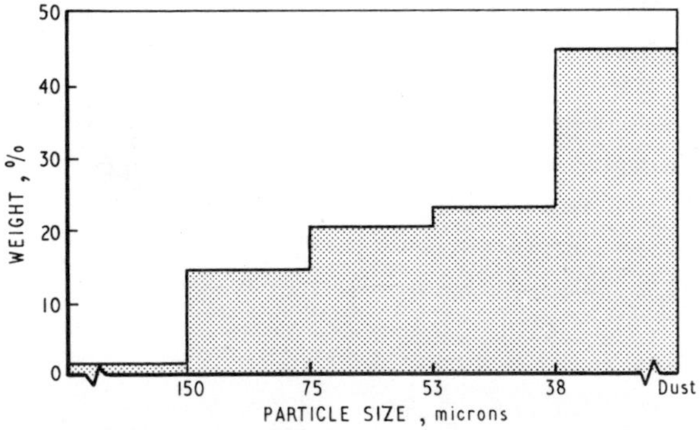

Fig. 17. Typical particle size distribution for water-atomised powders.

Fig. 18. The Hall flow meter. A simple and accurate method of measuring flowability of a metal powder. The powder, in this case stainless steel, can be seen falling into the receiving container on the base of the meter.

In direct powder rolling, the powder must flow freely, and any tendency to stick must be minimised. A common technique of measuring flowability is to time the flow of a given weight of powder through a standard orifice; this is the basis of the Hall Flow Test[27] (Fig. 18). Powders with a standard Hall flow meter rating of 30 sec/50 g are satisfactory. However, if the flow is excessive, the powder will flow through the roll gap without being fully compacted. Under these conditions, poor quality strip would be produced (Fig. 19). In many instances, this could be overcome by changing the flow characteristics of the powder, for example by increasing the moisture content.

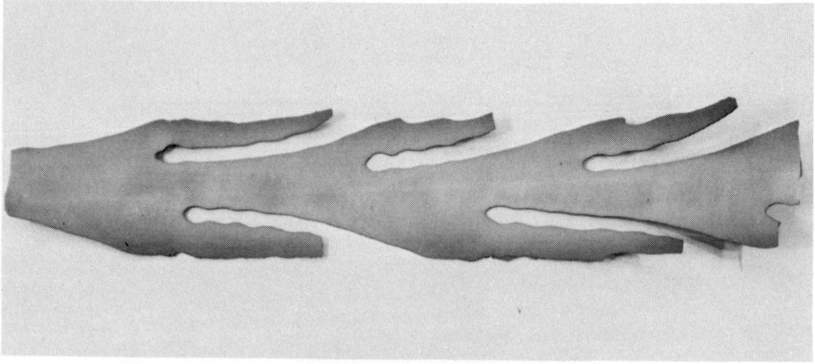

Fig. 19. "Fir tree strip". Produced by attempting to compact powder with excessive flow characteristics.

Particle Surface
To reduce the sintering time and increase the green strength, the oxide film of powder particles should be minimal. This is especially true in the case of stainless steels, where stable chromium oxide forms on the particles. This oxide is extremely difficult to reduce during further processing.

Roll Compaction
The direct roll compaction of metal powder into strip has been described by several workers. The main work described here relates to the research programme[28] carried out at the British Steel Corporation Sheffield laboratory using two-high horizontal mills, the first having

18 cm diameter × 12·75 cm wide rolls, and the second 61 cm diameter by 10 cm wide rolls. The principal aim was to produce compacted powder suitable for processing by the shortest possible route into strip possessing properties similar to those of conventionally rolled material. Studies showed that an unsaturated feed system led to considerable control difficulties with direct powder rolling. A simple saturated feed system also failed to produce the desired result, but was more stable and capable of providing acceptable strip at very low line speeds. As the head of powder in the roll bite decreased below a given value, the mass per unit area of the compacted strip also decreased.

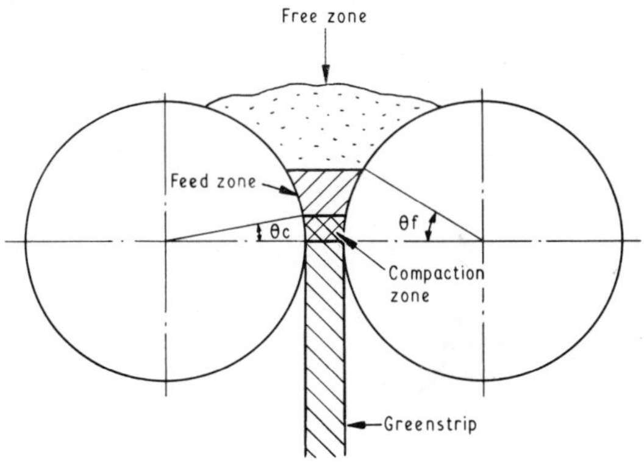

Fig. 20. Zones within the roll bite.

There were three distinct zones in the roll bite, as shown in Figure 20. These can be described as follows:

(1) A free zone, in which the powder falls under gravity; variations of powder head within this zone have negligible effect on strip properties.
(2) A feed zone in which the powder, though remaining non-coherent, is being dragged down by the rolls.
(3) A compaction zone, starting at what is commonly termed the friction angle, within which the powder has become coherent.

Consider first the feed zone; that is, the zone contained within the angle θf but outside θc. By an analysis similar to that used to determine the gripping angle in conventional rolling theory, it can be shown that $\theta f = \tan^{-\mu}$, where μ = the static coefficient of friction between the powder and the roll surface. The value of θf for various powders and roll surface finishes can be determined by simple static tests. Powder is poured onto the top of a static roll, and the angle of contact for which the powder slides off is measured (Fig. 21). From simple friction theory, we get $\gamma = \tan^{-\mu}$. Thus, for a given powder and a set of rolls, it is possible to calculate the maximum effective feed angle. This implies that it is possible to control the thickness of strip, for a given roll diameter, by shielding the roll surface to vary the feed angle.

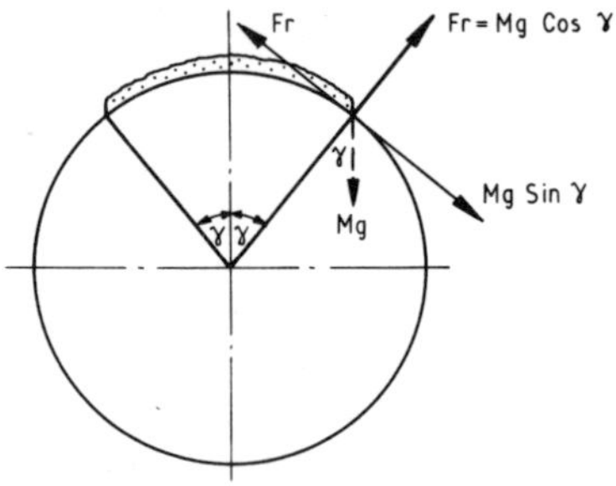

Fig. 21. Static determination of the limiting value of the contact angle.

The results of some experiments to determine the mass unit area of compacted strip as against shielding plate angle are shown in Figure 22. It can be seen that the angle at which movement of the shielding plate ceases to be effective correlates very well with the angle γ as measured above.

Not all the energy, in terms of the rolling load, is used in compacting powder into strip through the roll gap. A very small proportion is lost in squeezing metal out sideways (termed "spread") and along the direction of rolling (termed "elongation"), rather than in raising the density of the green strip.

It is known that small amounts of spread and elongation

occur towards the end of the compaction zone. However, if this is ignored, and the assumption made that only compaction takes place, a simple formula can be derived from the geometry of the roll pass which will closely approximate to the observed behaviour of the process, and will enable the density of the green strip produced to be easily determined:

$$\rho = \rho_a \left(1 - \frac{R\theta c^2)}{h}\right.$$

where ρ = green strip density, ρ_a = apparent density of powder as it enters the compaction zone, R = roll radius, h = the roll gap (strip thickness if no elastic recovery is assumed), and θc represents the angle of compaction. This relationship can be used to determine values of θc for different powders and roll surfaces. Most powders that have been successfully compacted into strip have indicated values of θc between 8 and 10°.

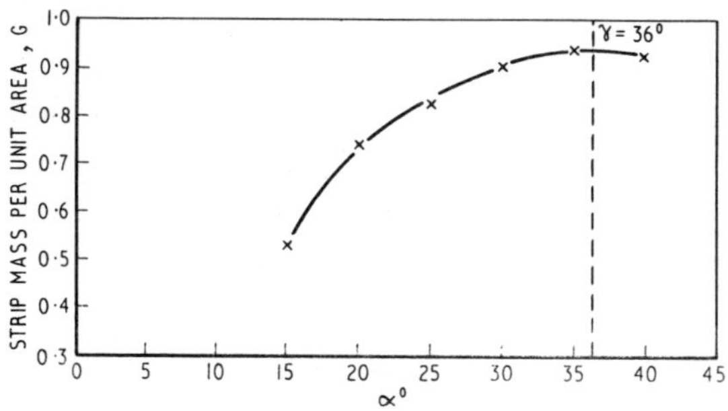

Fig. 22. Mass per unit area compacted strip v. shielding plate angle.

Process Control

In the initial experimental work, continuous independent control of density and thickness could not be obtained with either of the experimental mills, because the mill screw-down mechanisms would not operate under load. However, sufficient evidence was collected to show that strip could be produced over a range of thicknesses and densities.

A production powder rolling mill control system would probably take the form shown in Figure 23. In this system the mass per unit area of the strip is measured by radio-active techniques, and strip thickness by a mechanical measuring device. From these measurements, the density of the strip can readily be computed, and it becomes possible to design into the plant operation a closed loop system to control and monitor the green strip. A major

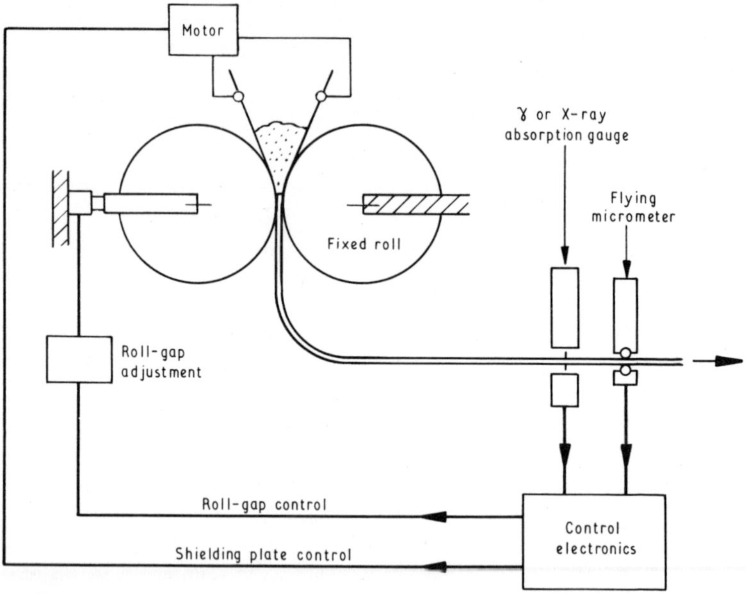

Fig. 23. Schematic diagram for a production strip from powder mill.

problem encountered in the direct powder rolling of strip prior to the development of the BISRA* direct powder rolling process, was the control of the transverse density distribution. With static hopper-side plates, the friction between the hopper wall and the powder reduces the powder feed to the edges of the roll. This effect coupled with loss through powder leakage results in a parabolic density distribution, as shown in Figure 24. Strip of this type has very weak edges, leading to serious handling problems, and consequent high yield losses while in the green state. Further processing after sintering is also

* Now BSC Corporate Laboratories.

difficult, owing to edge cracking caused by differential elongation during cold rolling to theoretical density.

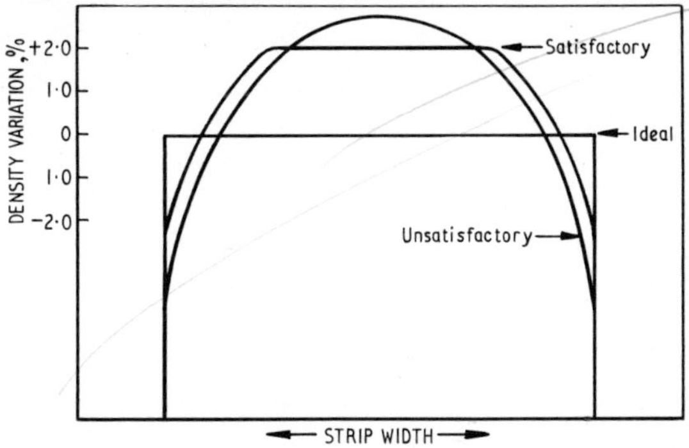

Fig. 24. Strip transverse density distribution.

To overcome these problems and assist in maintaining shape during subsequent rolling to finished gauge, the transverse density distribution must be maintained to close limits.

The forces involved in powder rolling are affected by both powder characteristics and mill parameters. These relationships are complex, and are not as yet fully understood. In general, the following trends have been observed:

(1) At any given thickness for a given roll size, rolling loads and torques increase as density increases;
(2) Similarly, at a given density, rolling torque, and to a lesser extent load, increase with increases in strip thickness;
(3) Reducing roll diameter from 61 cm to 19 cm reduces rolling load by approximately 50%;
(4) For a given roll diameter, there is a maximum strip thickness that can be compacted satisfactorily. It should be noted, however, that a wide range of strip thicknesses below this maximum can be compacted on a given roll size.

The relationship between green strip density, thickness, rolling load and torque has an important effect on the further processing, plant design and the overall economics of the process. In general, if the green strip density is kept low in the region of 70% theoretical, a light mill can be designed to compact the powder. However, this creates handling problems, and requires greater reductions during cold rolling to obtain the desired properties in the finished strip. Conversely, if strip with green densities of greater than 95% is produced, high rolling loads and torques are encountered at the compaction stage. The green density lying in the range 80–90% has been found to be the most suitable for further compaction. In this density range, the green strip can be handled effectively, and it does not blister during sintering. These points were borne in mind when the BSC 10 cm wide pilot production plant was constructed and commissioned at the Sheffield laboratories. This plant was set up to produce coils of sintered strip continuously from powder, which was then cold reduced on a standard mill available in the laboratory.

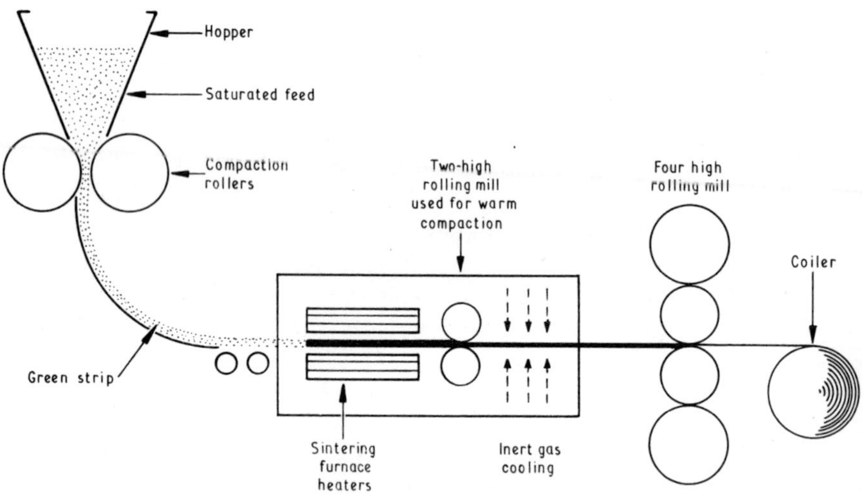

Fig. 25. Schematic layout for direct powder rolling using either hot or cold finishing. In hot finishing, final densification is done by the two-high mill within the furnace. Cold finishing is done through the four-high mill after the sintered strip has emerged from the furnace.

It can be seen from Figure 25 that dry gas-atomised powder is fed into the mill hopper, and compacted between the two 61 cm diameter hardened steel rolls. The compacted strip is then gently turned through 90° and fed into the sintering furnace. The sintered strip is then cooled under a protective atmosphere and coiled ready for further processing. A general view of the compaction mill and sintering furnace is shown in Figure 26.

Fig. 26. The compaction mill and sintering furnace of the 10 cm wide direct rolled strip from powder laboratory pilot plant at the BSC Hoyle Street Laboratory, Sheffield.

Sintering

The green powder-rolled strip has comparatively little strength and ductility, and must be subjected to certain mechanical and thermal treatments to obtain properties comparable with those of conventionally produced material. The method used for sintering can control the economics of the process and, in view of this, considerable effort is being made to improve the sintering techniques. The sintering furnace used at the Sheffield laboratories operates in line with the powder compaction mill.

Prior to the construction of the pilot experimental plant the precise nature of the sintering cycle required to provide strip with a quality and appearance equal to or closely matching that of conventional material was determined. A further aim was to reduce the furnace length to the minimum possible while allowing for high line speeds. For this reason, a short-time high-temperature sintering cycle was developed. Metallurgical studies to determine the optimum sintering cycle were carried out, using 150 micron-to-dust water-atomised type 304L stainless steel powder. Short lengths of compacted material were processed to finished strip, and properties checked against conventional material. A short time, high-temperature sintering cycle gave satisfactory properties in the finished material. The subsequent processing completed the densification and improved the mechanical properties and appearance of the sintered strip. The surface finish and corrosion-resistance of the strip must be closely comparable to those of conventionally produced material, or the available market for the finished strip will be severely restricted.

A processing cycle was developed to achieve the desired mechanical properties, and the effect of the process variables on corrosion resistance and surface finish was studied. Typical results obtained from experiments to determine the effects of process variables are shown in Figure 27 and are closely comparable with conventional material in most respects.

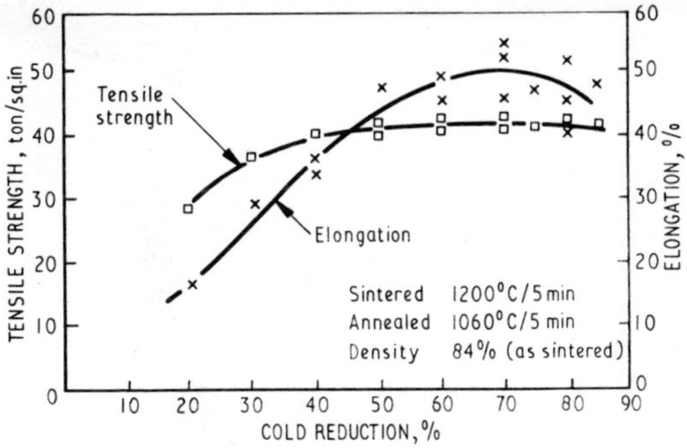

a. Sintering temperature 1200°C

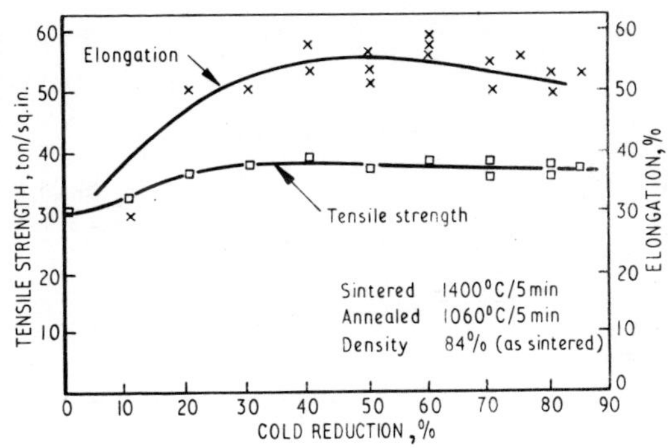

b. Sintering temperature 1400°C

Fig. 27. The effect of varying amounts of cold reduction on strip properties.

In all the treatments considered, the tensile strength rises to a maximum after a given level of reduction, and remains substantially constant as the amount of cold reduction is increased further. The elongation of strip produced from material sintered at 1200°C showed marked rise with increased cold reduction of up to 60–70%. Beyond this point, further reduction results in a decrease in ductility. In the case of 84–90% dense material, sintered at 1400°C, the ductility reaches its maximum value at a level of cold reduction lower than that sintered at 1200°C. The 84% dense material reaches its maximum ductility of 55%

elongation after 50% cold reduction, while the 90% dense material has a ductility of 60% elongation after 40% cold work.

In all the combinations of treatment considered during the experimental programme, after a certain level of cold reduction the ductility shows a tendency to decrease with further cold work. This was originally thought to be due to changes in the physical dimensions of the test piece, but this was only a partial explanation; further investigations have shown that part of the deterioration can be accounted for by mechanical damage to the metallurgical structure resulting from overworking.

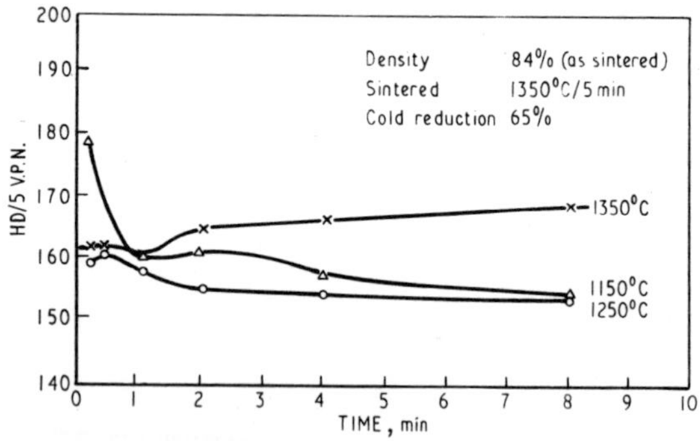

Fig. 28. Variation in strip hardness with annealing time at three temperatures.

Effects of final annealing time and temperature on the properties of cold rolled strip have been determined for products starting with a range of green densities from 70–90% theoretical. Typical results, showing the variation of strip hardness with annealing time for temperatures of 1150°, 1250° and 1350°, are shown in Figure 28. Annealing temperatures higher than those used in conventional processes appear to be beneficial. A typical microstructure of the finished product can be seen in Figure 29.

It is interesting to note that annealing at 1350°C causes a very rapid initial softening followed by a steady increase in hardness with time. This is due to the formation of ferrite in the matrix. In general, the number of anneals given to samples during cold rolling has very little effect, if any, on the mechanical properties for reductions in thickness of less than 80%. Above this reduction level, slight increases in elongation have been observed in specimens given two intermediate and a final anneal. However, it has subsequently been found that by closely controlling the initial sintering cycle and cold rolling operation, it may be possible for certain applications to dispense with an annealing operation altogether.

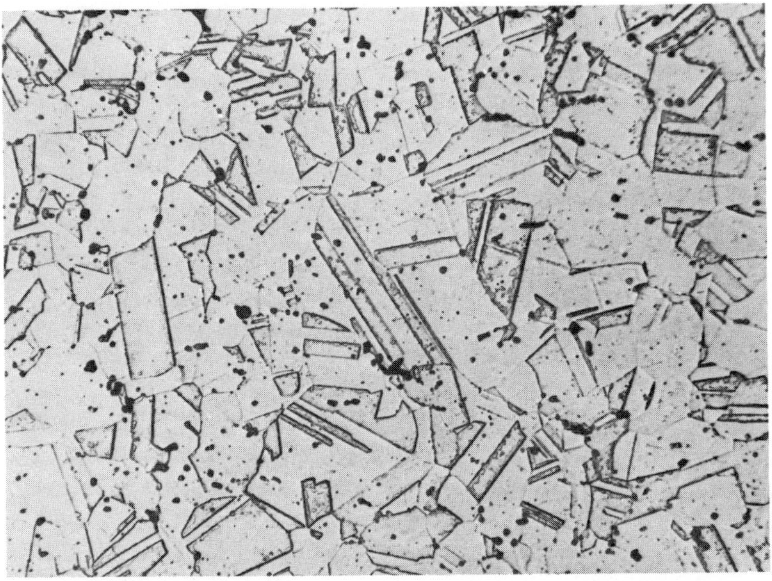

Fig. 29. Typical microstructure of powder-rolled stainless steel strip.

Evaluation of the BSC-BISRA Direct Powder Rolling Process

Evaluation of the BSC-BISRA direct powder rolling process for manufacturing stainless steel and alloy products indicates a processing cost for a powder route which may be substantially less than that for similar material produced conventionally. Such economies result primarily from the reduction in overall yield loss, which may amount to less than 10% compared with losses in the region of 30% for conventional processing, together with the elimination of processing and capital costs associated with reducing the ingot and slab to hot band. It is also clear that these cost advantages become increasingly significant as the gauge of the finished product is reduced. The product, quality holds out high promise. The mechanical properties are similar in all respects to those for conventional material. Surface quality is satisfactory, except for the very high standards required for certain applications, e.g. pharmaceutical and medical requirements. (These represent only a small fraction of the total market.) Work on corrosion indicates that a satisfactory standard can be achieved. Furthermore, the development potential of the direct powder rolling process is by no means exhausted. Work is continuing to widen the range of strip thicknesses that can be produced by a given piece of plant. Although the TSP (Thin Strip Process) process described later offers an attractive route for making foil and the thinnest gauges of materials, in stainless steel, mild steel and non-ferrous alloys, the attractiveness of the strip from powder route would be increased by having a direct powder rolling plant which could roll in the size range from 4 mm to 0·3 mm finished gauge.

The BSC-BISRA powder rolling process does not differ in basic principle from the more specialised small-scale applications already commercially exploited. The principal fact is that the restriction to low rolling speeds and the material limitations which have to be accepted with the previously operated processes have been overcome. The developments lie almost entirely in the field of having a better understanding of the process variables and their control. This new ability now makes it possible to apply a direct powder rolling technique to the commercial production of considerable tonnages of metal strip.

It is to be expected that the next stage of commercial development will now be undertaken and that during the mid 1970s a reasonable percentage of the stainless steel strip marketed in the thinner gauge range in the U.K. will be manufactured by the powder route.

SLURRY PROCESSES

From time to time, suggestions have been made for the production of strip from powder by methods which have derived their inspiration from the process known as slip casting. In slip casting, a metal powder is mixed with a suitable solvent and organic binder, cast to shape, and allowed to set; the binder and solvent are then burned off during a heat treatment operation. The majority of the proposals (e.g. that of the Kennecott Copper Corporation[29]) incorporate a complex organic binding agent, usually rather expensive, to deposit and form a film which is subsequently sintered and cold rolled.

The use of expensive solvents and binding agents added greatly to the cost, and the majority of processes[30] produced discrete lengths of material, rather than continuous strip which were sintered to give material of low density before roll compaction. Thus, the amount of work that can be done without additional heat treatments is limited by the low strength of the particle bonds developed during sintering. In consequence, a great deal of cold rolling and heat treatment is necessary to achieve reasonable properties in the final product.

A METHOD OF MILL FEED FOR HIGH SPEED COMPACTION

The compaction of powder is generally regarded as the first stage in the production of strip by direct powder rolling, and in spite of the developments made in process control it is not possible to envisage the direct cold powder compaction being carried out at a speed remotely approaching that of a conventional rolling mill producing black plate. Because of this, it has been supposed that the most economical route would be to make dense, thick green strip, sinter it, and then cold roll it to the finish gauge, and this in fact was the principle on which [illegible] of the slip casting techniques based their approac[illegible] ever, as the gauge is reduced by rolling from the initial thick green strip, the price of the material increases.

It has been suggested that the best answer would be to powder-roll directly into thin strip, so that this would need only the minimum further reduction to final gauge consistent with the strip properties required. It is assumed, of course, that the process would be suitable for large tonnage production; that is, production far greater than anything envisaged for the medium gauge stainless steel powder process developed at the BSC Corporate Laboratories, Sheffield.

The process developed[31, 32] at the BSC Sketty Hall Laboratory, Swansea, employs a film-forming technique to overcome the problems encountered in metering loose powder feed to the compaction mill.

The basic steps of the process are the initial mixing of iron or other metal powder with a suitable binder and water, to form a slurry. This is spread on a temporary substrate, such as a metal strip or belt, to a desired width and thickness and dried to form a self-supporting film. The film is peeled from the substrate and then cold roll-compacted to form a green strip; the binder is then burnt off and the strip is sintered in a reducing atmosphere, and then cold rolled. A further sintering operation follows, also in a reducing atmosphere, and the strip is then finished by a planish or temper rolling operation. It has been shown that the process can make acceptable mild steel strip using simpler and cheaper plant, and if necessary at lower levels of output, than the conventional route.

Thin Strip Process

4. The TSP Process

The development of the film-forming approach to feeding a compaction mill followed several years' research on powder coating processes for steel strip. In clearly experimental work on the development of the aluminium powder coating process, electrostatic deposition techniques were used to make the coating adhere to the steel substrate prior to compaction. It was discovered that, under certain processing conditions, it was possible to produce a coherent aluminium powder deposit, which could be peeled off the steel substrate after roll compaction to give a green strip of aluminium. Samples subjected to a sintering operation showed quite good properties. It was thought that this technique might open the way to the development of an unsaturated feed system, having no free powder in the mill bite. The rate of compaction could then be determined by a metering device independent of the mill, feeding the powder into the rolls and allowing the production of compacted strip over a range of precise gauges.

Sketty Hall also had other factors to consider. The speed of compaction had to be an order of magnitude greater than the previous best, implying rolling speeds of the order of several metres per second. For maximum saving on capital and operating costs, only a relatively small degree of rolling could be tolerated to develop the required thin strip properties. Sintering time had to be very short to allow for a practical in-line sintering furnace.

The technique of coating and compacting upon a temporary substrate, followed by peeling-off of the green strip, appeared to offer attractive prospects in respect of these requirements. However, it was soon found that the compaction of a film upon a substrate in the case of a permanent re-usable substrate, led to the rapid deterioration of material from which the substrate was made, whereas the cost of an expendable substrate severely affected the economics of strip production. Further investigation was carried out, therefore, leading eventually to the slurry process. This enabled the powder to be metered accurately to the mill in the form of a self-supporting iron powder/binder film, which could be peeled from the substrate before compaction.

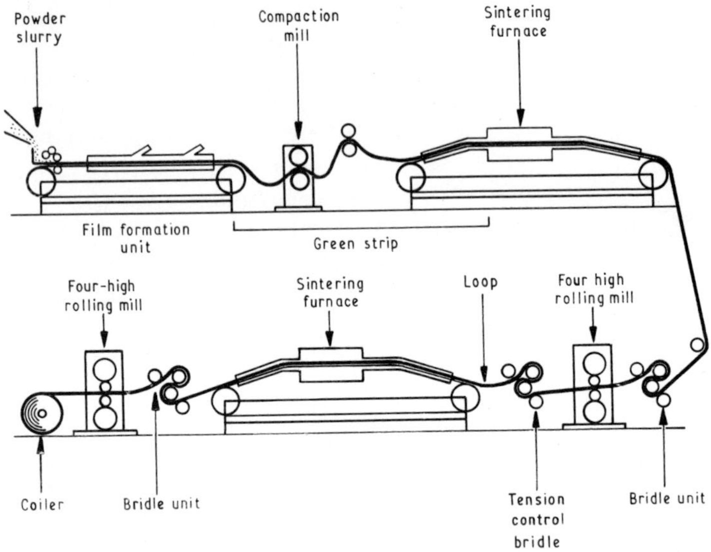

Fig. 30. Probable layout of an industrial TSP line.

This technique formed the basis of the TSP process, which can be used to make strip from a wide range of metals and alloys. The schematic diagram (Fig. 30) shows a possible configuration of an industrial line for making strip from powder by the TSP process. Figure 31 compares the TSP process route from hot metal to finished strip with a conventional route (simplified for the sake of clarity).

THE SLURRY, DEPOSITION AND FILM FORMATION

The first stage in the TSP process is the formation of a homogeneous, aqueous slurry of iron powder and a film-forming binder. The binder must be usable in aqueous solutions at low concentration, and provide an effective

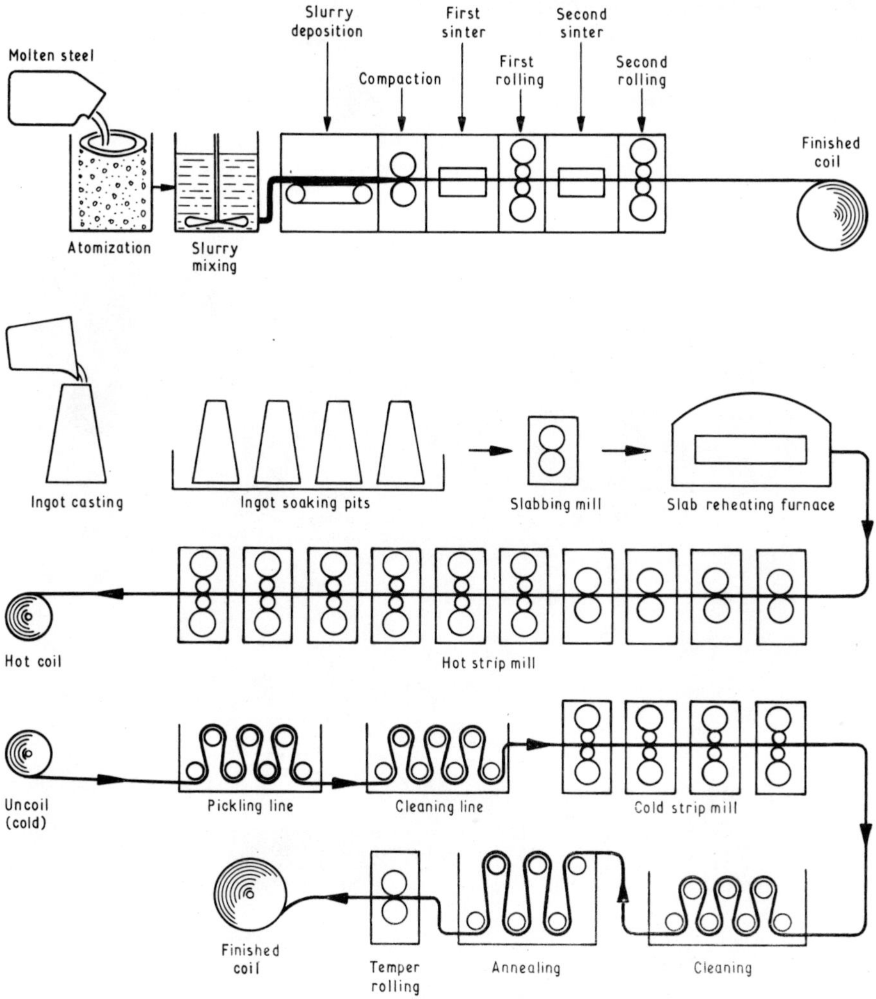

Fig. 31. TSP route and conventional rolling compared.

suspension of powder. It had to be available in a wide range of viscosity grades, form a coherent and self-supporting film when dried, and be readily driven off during sintering operations. It was desirable that the binder, during compaction, should act as a lubricant and so promote closer particle packing to maximise the density of the compacted strip.

A typical slurry formulation that has given consistent results for mild steel strip is 70% by weight of metal powder, 29·4% water, and 0·6% water-soluble binder. Several types of iron powder have been used successfully, and the viscosity of the slurry can be varied between 100,000 and 1,000 centipoises to suit the method of deposition.

Fig. 32(a). Film formation unit on the original Sketty Hall Laboratory (Swansea) research plant used roller coating techniques to deposit the slurry.

In commercial use, the TSP process would demand the laying down of consistent wet coatings at speeds of 50–250 m/min. Roller coating methods, similar to that shown in Figure 32(a), have been and are being studied, and other methods such as curtain coating (Figure 32(b)) are also under investigation. Slurry viscosities suitable for roller coating lie in the 1,000–5,000 centipoise range; at the lower viscosities it is necessary to stir the slurry constantly to prevent powder settling out. To ensure accurate coating, the substrate must have consistent dimensions, good shape and an acceptable surface finish.

Fig. 32(b). A curtain coating unit developed for the deposition of iron slurry uniformly at high line speeds and commercial strip width.

The TSP pilot experimental unit built at Sketty Hall consisted of a coating unit fitted with a stainless steel band on which the slurry was deposited by a four-roll reverse roller coater. A small doctor roll, mounted above the pick-up roll, served to reduce ribbing effects. Each roll was equipped with its own variable speed drive, thus permitting a wide range of coating thicknesses and application speeds to be used to continuously deposit slurry coatings in the range of $\frac{1}{3}$–1 mm thick.

The deposited slurry film can be dried in about 30 seconds, depending on its thickness. On the laboratory unit a bank of quartz halogen heaters, located below the band, were used for this purpose. Two fans situated above the band blow warm or cold air along the surface of the slurry film, removing water vapour and preventing boiling of the binder.

FILM COMPACTION TO "GREEN" STRIP

The removal of the dried film from the substrate is assisted by a slight contraction which occurs while drying; this breaks any temporary bond of the film to the band. When removed from the supporting band, the resulting iron powder binder film is flexible and quite strong, and may be readily handled during transfer to a compacting mill. Green strip up to $\frac{1}{4}$ m wide has been compacted continuously on the laboratory plant and short lengths of film have also been handled and compacted successfully at speeds ranging from 15–155 m/min.

There is an extensive background of experience in deposition technology in respect of paints and organic coatings similar in their general characteristics to the metal powder slurry. Information available suggests that the continuous in-line operation of the TSP process, at least in respect of its deposition and compaction stages, can be carried out at speeds in the region of 150 m/min.

SINTERING THE "GREEN" STRIP; ROLLING AND FINISHING

In the laboratory pilot plant, sintering is carried out in a 45 kW electrically heated 3-zone furnace of humpback muffle type. The rectangular muffle measures 0·05 by 0·31 m internally. A nitrogen and hydrogen atmosphere is supplied to the furnace through a control and mixing unit, the gases being burned off at the top of two vertical tubes a short distance in from each nitrogen-purged box. Through the furnace runs a continuous mesh belt, woven from 80–20 nickel–chromium wire, which carries the front edge of the strip issuing from the compaction mill through to be coiled or passed to a second mill for a further sintering or rolling operation.

The sintering stage has been simulated in line for gauges of strip up to 0·25 mm, and samples approximate to this furnace temperature in 7–11 seconds. Once the binder is

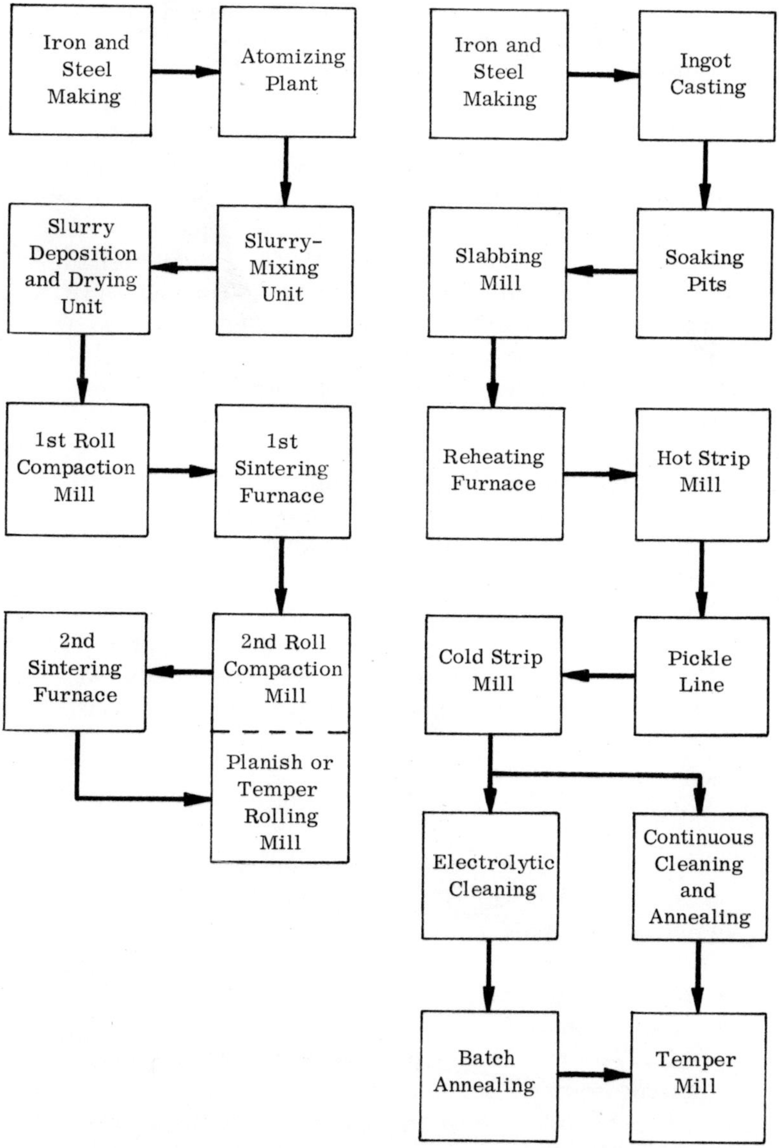

Fig. 33. Mild steel strip from powder route compared with that from black plate production (can stock).

burned off, sintering results in a strip with a tensile strength of 154 MN/m^2, having a tensile elongation up to 1% and about 90% of theoretical density.

The three final stages of the process route start with the rolling of the green sintered strip to give a nominal extension in strip length, to a fully dense material. A second sinter may then be carried out to further improve the strip strength and ductility; a final temper or planishing roll is then carried out to improve the strip's shape, remove minor surface blemishes, impart the required surface finish, and (depending upon the product) further change the mechanical properties appropriate to the desired strip quality. It is interesting to note the reduction in the number of process stages between the conventional and powder route shown in Figure 33 for the production of black plate.

On the experimental unit at Swansea the final stages of strip rolling were done with a four-high mill, having $0{\cdot}13 \times 0{\cdot}3$ m work rolls, and $0{\cdot}3 \times 0{\cdot}3$ m back-up rolls. Commercial plant will almost certainly be designed to perform all seven stages of the process on a continuous in-line basis. As in the case of direct powder rolling, the commercial success of any new process rests finally on its economic acceptability and the quality of its products.

COMMERCIAL APPLICATIONS

Although the TSP process was conceived initially as a means of making mild steel strip for use as can stock material, it seems almost certain that the large scale tonnage application of the process to this market must await the completion of further research aimed at reducing the cost of manufacturing large tonnages of mild steel powder. This may seem somewhat disappointing, as metallographic examination of mild steel strip samples produced by the TSP route show that, apart from the absence of carbides, the structures resemble conventionally produced strip; the process thus seems capable of yielding a material equivalent to conventionally produced thin mild steel strip for the manufacture of tinplate. It must be noted, however, that the suitability of strip for tinplate application is influenced by other factors such as ageing, work-hardening, ductility, ultimate stress, yield stress and corrosion resistance. Extensive testing of strip made from powder has therefore been continuing, and while the general effects of processing conditions on the final strip properties are known, it is only now becoming possible to determine the range of tinplate tempers that can be produced by the powder route.

FUTURE OF THE TSP PROCESS

In the authors' view, the TSP process will almost certainly find its first applications in the manufacture of thin stainless steel foil, in the thickness ranges below the practical minimum mentioned in connection with the process developed at Sheffield. At Swansea, initial trials in making stainless steel foil were conducted with water-atomised 300 mesh type 304L stainless steel powder. In general, stainless steel foil produced from powder shows good tensile strength and lower tensile elongation than conventionally produced material. This resulted at least in part from the dispersion hardening by oxides not reduced during the sintering process. As a result of further research, it is now possible to control this factor more closely by varying the processing conditions, and to indicate optimum conditions to achieve strip properties suitable for particular commercial requirements.

Outdoor corrosion trials on stainless steel samples have been in progress for some time, and are generally satisfactory. The economics of the TSP process reflect its freedom from the main limitations normally suffered by powder rolling techniques. The effective metering system gives an accurate powder distribution well upstream of the roll nip. The compacted strip thickness is controlled by the wet slurry deposit, and does not depend on roll geometry, friction at the roll surfaces or the flow characteristics of the powder. The inclusion of the binder permits compaction speeds as high as several metres per second without disturbing the incoming powder. For the production of thin strip from iron powder, the process lends itself to short sintering times and low total rolling reductions. Moreover, the process is technically capable of making strip from many other metals. The largest use of thin strip is in the manufacture of mild steel black plate for conversion into tin plate, and over one million tons is produced each year in the United Kingdom, mainly in the gauge range of 0·2–0·3 mm, with a general trend towards thinner materials.

The original main objective of the TSP process was to produce strip suitable for the manufacture of tin plate or tin-free can stock, more economically than by conventional means. The suitability of the TSP strip process for this purpose is shown by the fact that sample materials have been coated successfully with tin and with Hi-top (a chromium–chromium oxide coating) by the normal electrolytic method, but the critical factor is selling price. Here the price of mild steel powder is far more crucial than that of stainless steel, and the economic assessment of black plate production by a powder metallurgical route begins and ends with the cost of the powder.

ECONOMICS OF THE TSP PROCESS

Calculations indicate that mild steel strip can be made from powder at least as cheaply as today, while saving approximately 30% on capital cost, providing a base powder price can be achieved equal to the metal cost at teeming, plus an atomisation charge of the order of 22–33 dollars per ton suggested by American workers[33] and the lowest possible additional handling charges.

The main advantage of the TSP process for making black plate, therefore, is a better return on capital investment, and a secondary advantage is that economic units to manufacture $\frac{1}{4}$ mm black plate in increments of 125,000 tons per annum are possible, as compared with a minimum of something like 500,000 tons/p.a. for the conventional route. Thus, although the TSP process is technically ready for the final stage of development into a full-scale production process, its exploitation must be considered as a longer-term objective because of the relatively large quantities of powder required at low cost for the commercial success of the process. However, a number of shorter-term objectives exist, the most immediate of which is the production, at a line speed in the region of 6 m/min, of a premium product such as thin stainless steel foil from a pre-alloyed powder. The cost of producing 0·2–0·125 mm strip in 304L stainless steel at speeds of 1·5–12 m/min, assuming 95% yield, has been calculated.

At an output level of 6 m/min, and on the basis of three shifts per day, the total strip produced would be of the order of 800 tons/p.a. At this level of output, the powder price could fall to something like half the present-day (1972) market price. Comparison with current selling prices of stainless steel strip indicates that powder price is the dominant factor in the total processing cost at line speeds greater than 12 m/min.

It is worth noting that the production of extremely thin foil, say of 0·025 to 0·10 mm finished thickness is inherently a very expensive process by the conventional route, and the cost of the finished product is such that its full market potential has never been available for full exploitation. The production of thin stainless steel foil by the TSP route could change this pattern markedly.

5. Direct Compaction of Hot Powder

The economics of powder metallurgical routes are attractive for moderate gauges down to foil in the case of stainless steel, and for thin (can stock) mild steel, but techniques involving the compaction of cold powders become economically less attractive as the finished gauge of the required strip is increased. Although the upper gauge limit is dependant on the metal involved, there is, however, a no-man's-land between those strip gauges which can be produced economically by the cold compaction of powder and the thicker materials which, in the case of mild steel, are required in considerable annual tonnages.

Spray rolling may be one possible production route, but the problems of control are considerable. There is, however, a powder metallurgical process which can avoid the limitations which are faced by cold powder compaction. This involves the hot compaction of an extremely coarse powder, in which individual particles may be likened in size to shot or rice grains.

REYNOLDS METALS; CENTRIFUGAL ATOMISATION; THE COMPACTED SHEET PROCESS

A process based on the hot compaction of coarse powder has been successfully operated by Reynolds Metals(25) for the production of aluminium strip, including aluminium alloys containing up to 12% magnesium (Fig. 34). Coils of finished strip up to 1 m wide, weighing up to 11 tons, have been manufactured, and the properties of the finished strip have been extremely satisfactory. In the case of aluminium, however, the inducement to adopt such a process did not arise from the need to achieve very high outputs of thicker strip. Rather, Reynolds Metals sought to obtain the benefits of the powder metallurgical route--

such as the reduced capital requirements—by the elimination of much of the rolling required to produce finished strip, while at the same time avoiding or largely reducing the amount of oxide which would be included in the product. As practised by Reynolds Metals, the basic process consists of rolling rice-grain-sized powder particles produced by centrifugally spinning a perforated cylinder into which hot metal is introduced. This part of the operation is carried out under an inert atmosphere, usually nitrogen, and the powder is conveyed while hot through a hopper system into a compaction mill, where it is directly rolled to strip in one pass at speeds exceeding 60 m/min. This strip is fully densified and metallurgically bonded and no further sintering or heat treatment is required.

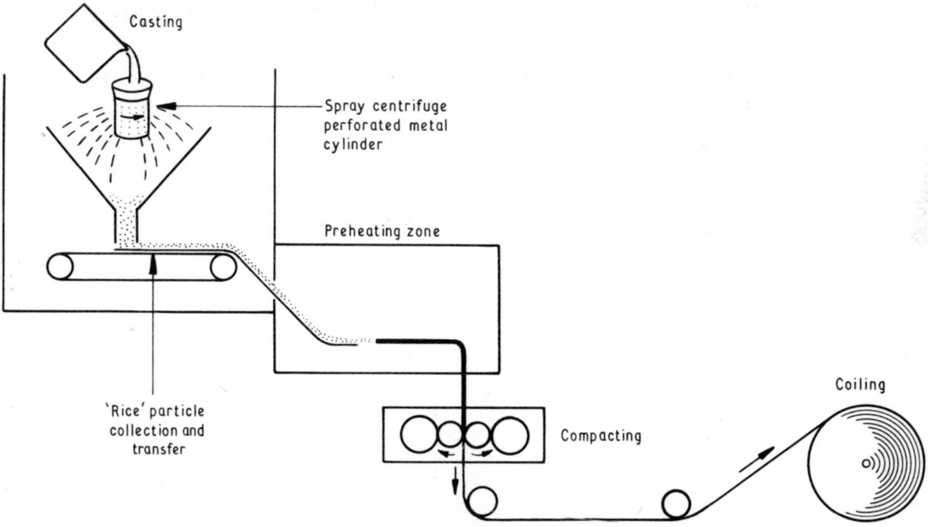

Fig. 34. The Reynolds Metals compacted sheet process. Consists of centrifugal atomisation, "rice grain" powder collection, transfer through a preheating zone to achieve the desired powder temperature, into the roll nip for direct hot compaction into fully dense finished strip in a single pass.

Although the concept is elegant, and the plant sophisticated, there are a number of factors which would make such a route impractical for the bulk production of mild steel strip at a price competitive with the present-day conventional routes. Firstly, the operating temperature of the process would have to be raised from around 660°C to the region of 1,540°C; secondly, the powder atomisation system would be quite incapable of producing the quantity of powder required, or in fact having a reasonable life expectancy under such rigorous conditions.

STEEL STRIP BY HOT POWDER ROLLING

Depending on the gauge of strip manufactured, a conventional plant for producing steel strip is required to make of the order of 1–4 million tons per year at speeds of 3–600 m/min, the strip widths lying in the range of 1–2·5 m. To achieve such a throughput by hot powder rolling, the process route would have to be virtually continuous, and Figure 35 shows the route an idealised process flow would probably take. Starting from liquid metal, atomisation into coarse shot similar to that shown in Figure 36 would

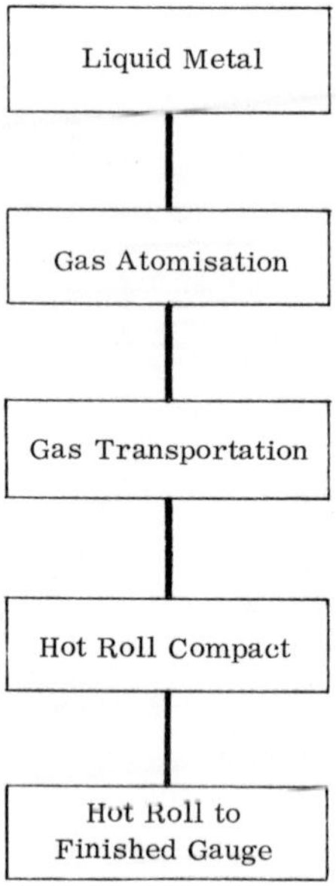

Fig. 35. Hot powder rolling process chart.

be carried out using an inert gas under an inert atmosphere. It would be necessary to use hot gas fluidisation, for example, to move the hot powder. The metal particles would have to be presented for compaction at the right temperature, necessitating accurate temperature control. A conventional compaction mill would process the hot particles into compacted strip; this would not require further sintering, but could then follow a conventional processing route involving rolling and annealing as necessary to produce the final coils of strip of the desired thickness. Apart from research done in the U.K., there have been two major attempts to develop and operate processes similar to that described above. It is fair to say, however, that none of these processes under development has as yet been operated in an integrated manner.

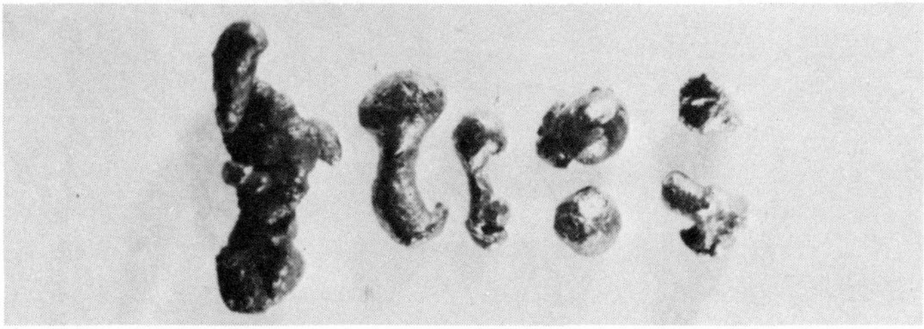

Fig. 36. Typical mild steel particles. Mag. × 3.

In America, the Republic Steel Corporation[26] has carried out considerable work on the development of a hot powder rolling route as an economic alternative to the production of strip by orthodox strip mills. In this process, hot coarse powder particles are delivered straight to the rolling mill, where they are compacted at speeds of up to 30 m/min (Fig. 37). Using this technique, it has been possible to produce strip with properties varying over a wide range, and deep-drawing steels suitable for the manufacture of car bodies, etc., have been satisfactorily produced.

It is interesting to note that the main argument for developing a hot powder rolling process for bulk tonnage manufacture of mild steel strip is one of financial investment.

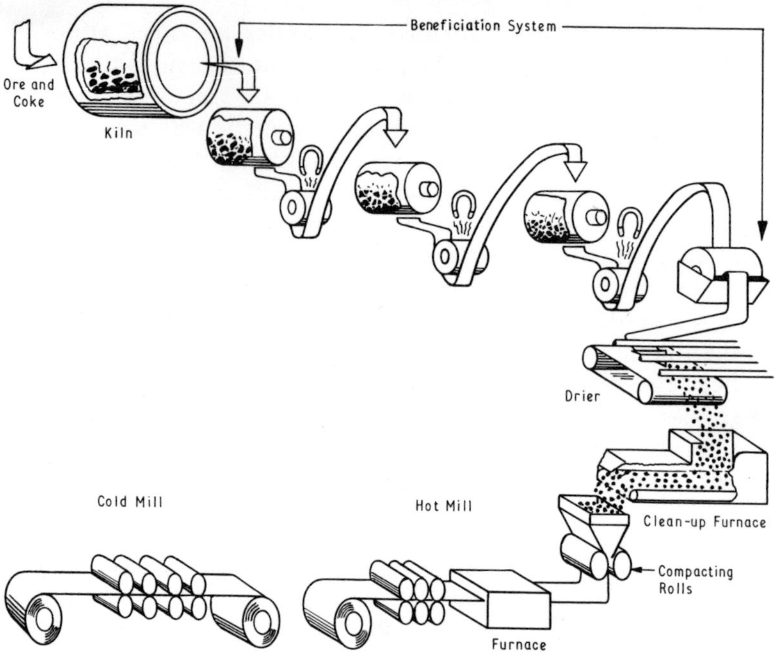

Fig. 37. Hot compaction of direct reduced powder—Republic 1962. (Iron and Steel Engineer, *February 1962, pp. 72–76.*)

The capital required to set up future generations of orthodox strip mills will be so high that re-capitalisation of any steel company will become increasingly difficult. On the other hand, hot powder rolling strip mills can be constructed at considerably lower capital cost. From the research work done so far it has been estimated, for example, that the total power installed in terms of electrical equipment would be reduced by a factor of 10 for the hot powder rolling route, compared with an equivalent conventional installation. At the present time, it is not known exactly when the Republic Steel Corporation will decide that sufficient research and development work has been done, and that the economic climate is satisfactory for it to invest in hot powder rolling as a means of strip production, but such a development may not be very long delayed.

Whereas research on the direct production of strip by hot

powder compaction is being done by a steel producer in the U.S., research in Western Europe has been carried out by an industrial plant builder: the Schloemann Company of Düsseldorf, West Germany. In the Schloemann process, powder in the form of steel shot is heated to 950°C in a controlled atmosphere. The diameter of the shot is approximately equal to the finished strip thickness (2 mm). Compaction of the hot shot is performed in a conventional powder compaction mill, and the subsequent processing of the powder rolled strip is carried out conventionally. During the research programme, hot powder rolled strip has been subjected to further cold reduction down to thicknesses of 1 mm; this material has satisfactory properties, and has been deep-drawn experimentally without any sign of rupture. The size and shape of the shot used in the process is not critical, and the ability to produce strip from a very crude raw material, basically so much cheaper than fine powder, will make the process viable over a range of strip thicknesses not normally attractive for a powder route.

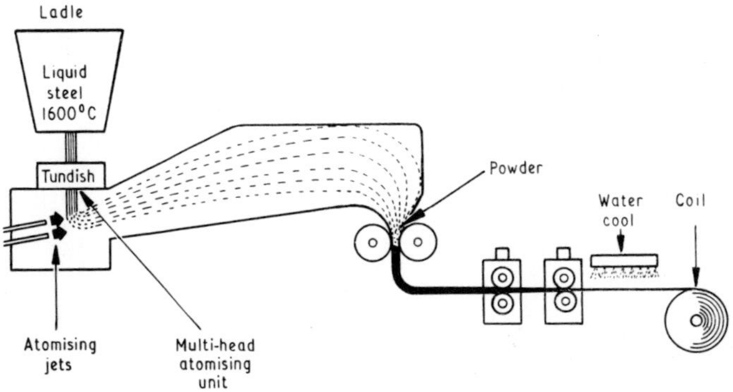

Fig. 38. Schematic diagram of the hot powder-rolling process.

Neither of the processes for the manufacture of steel from hot powder rolling described above makes any pretence of being an integrated process flow route at the present time. Both start by using relatively coarse powder particles, which may be either directly hot rolled from production, or re-heated immediately prior to compaction. In the authors' view, for such a hot powder rolling process to become a major commercial route for strip manufacture, it will be desirable to use as wide a range of particle size as possible for the hot powder feedstock. In this way, it will become possible to handle and transport the hot powder in a fluidised condition to produce feedstock for further processing. Moreover the economics of the process is likely to be correct, owing to the high yield of usable powder. The schematic diagram, Figure 38, illustrates the simplicity of a hot powder atomisation-transport-compaction process route.

6. Summary and Conclusions

A number of the processes described in this book have ceased to be laboratory curiosities, but have not yet had a major impact on the economics and technology of the metallurgical manufacturing industry. It is accepted that powder metallurgical processes may be used successfully in the case of exotic materials such as tungsten and other refractory metals. Is it likely, however, that powder metallurgical processes could ever become serious competitors to the conventional orthodox rolling mill as a means of strip production?

In general terms, the time lag between the initial conception of a new process or scientific discovery and its practical application is increasing rather than decreasing. The more complicated and the more capitally intensive the development, the longer the development period will be, particularly if the product or process is at the frontier of its particular technology; a new aircraft like Concorde is an obvious example. In the authors' view, it seems likely that the cold compaction processes for producing medium and thin gauge stainless steel and thin gauges of mild steel (Fig. 39(a) and (b)) will develop significantly during the early 1970s, and that the commercial manufacture of stainless steel and stainless steel foil by powder routes could commence during the decade. The possibility of producing large tonnages of steel or any other metal strip by a hot powder compaction route will be the subject of considerable research in the coming years, and it is possible that one or more of the process routes described may enter pilot commercial operation by the beginning of the second half of the decade.

The advantages to be derived from the manufacture of strip from powder can be summarised as follows:

(1) A considerable reduction in the size of plant

(a) Mild steel strip produced by the TSP route demonstrates that the material is suitable for normal forming techniques. Most of the mild steel tonnage made by the TSP process is likely to serve as the basis of tinplate, which in turn will be used to make various types of containers and can ends. In the background are vehicle radiator segments.

(b) Thin stainless steel strip produced from powder. These metal bellows demonstrate the good formability of the materials and show that it can be as satisfactorily welded as conventionally produced strip.

Fig. 39. Test items manufactured from strip from powder.

required to manufacture a given tonnage and thickness of strip, and hence a far lower capital investment required per unit of production.

(2) In the case of cold powder compaction routes, the ability of the manufacturer to respond to market developments at lower incremental output levels, thus keeping supply more closely in tune with demand.

(3) The ability to produce strip in widths and thicknesses not possible by the conventional route.

In respect of the third point, the exploitation of the new processes in ways not open to traditional techniques offers considerable potential. For example, because of relatively low rolling loads, it may become a practical possibility to use the slurry process for the production of mild and stainless steel foils in widths of up to 3·5 m, normally associated with papermaking. Metallic laminates could be produced by a combination of films at a suitable stage in this process. In the case of the cold compaction routes, there is the possibility of producing metal strip reinforced by strong, stiff filaments or whiskers.[34, 35] A logical development would be the use of steel foils and strips in combination with plastics and paper and other materials,[36] in the form of composites. Corrugated thin stainless steel foil on either side of expanded polyurethane foam would, for example, provide an excellent structural material for buildings, vehicles and freight containers. In addition to being resistant to abrasion, puncture, weathering and corrosion, it would be light and strong, and offer good thermal insulation.

Even on the more mundane level of continuing to develop existing markets, the growth potential for lower cost stainless steel is considerable. At the present time, the cost of rolling thinner gauges of material is such that it is often cheaper for a manufacturer to use a thicker gauge of stainless steel than is adequate for his purpose; further roll reduction, far from saving him money, would maintain or even increase his costs. The ability to produce a range of stainless steel strip, therefore, in which the price is more directly related to the quantity of material supplied, will see a real and significant growth in the usage of stainless steel. In the authors' view, this can only be achieved by the adoption of powder metallurgical routes for strip manufacture.

Finally, the use of powder techniques for more expensive materials offers great potential. The manufacture of aluminium and aluminium alloys, copper, copper alloys, nickel and refractory metals, is either already under way, or not far distant. Each metal presents its own particular development problems. None of these problems, however, is likely to constitute a fundamental barrier to the adoption of a powder metallurgical route, and it is therefore more likely that the final decision will be made entirely on the relevant economics for the production of the material concerned. The next decade will be crucial in the development and diversification of powder metallurgical routes for the production of metal strip, and it is one which all those concerned with the metallurgical industries will watch with great and growing interest.

BIBLIOGRAPHY

Goetzel, C. G., "Treatise on Powder Metallurgy", Vol. 3 (1962), Vol. 4 (1963), New York and London (Interscience Publishers).

Hausner, H. H. (Editor), "Modern Developments in Powder Metallurgy", Vol. 3 (1966), Vol. 4 (1971), New York (Plenum Press).

Hausner, H. H., Roll, K. H. and Johnson, P. K. (Editors), "Perspectives in Powder Metallurgy", Vol. 1, New York (Plenum Press).

Jones, W. D. "Fundamental Principles of Powder Metallurgy", London (Edward Arnold), 1960.

Leszynski, W. (Editor), "Powder Metallurgy", New York and London (Interscience Publishers), 1961.

Sands, P. C. and Shakespeare, C. R., "Powder Metallurgy", London (G. Newnes Ltd), 1966.

REFERENCES

(1) McManus, G. J., *Iron Age*, **205**, No 13 (March 1970), 73.
(2) British Patent 1,229810.
(3) British Iron and Steel Research Association, Annual Report, 1969.
(4) British Patent 1,218365.
(5) Singer, A. R. E., *Metals and Materials* **4**, No. 6 (June 1970), 246.
(6) Bessemer, Sir Henry, "An Autobiography" (1905), pp. 67–68.
(7) German Patent 154998.
(8) Naeser, G. and Zirm, F., *Stahl u Eisen*, **70** (1950), 955; also *Metallurgical Reviews*, No. 14 (1959), 179.
(9) Storcheim, S., Nylin, J. and Sprissler, B., US Atomic Energy Commission, Publ SEP 161 (1954).
(10) Hunt, D. G. and Eborall, R., *Powder Met.*, **S** 1 (1960).
(11) Crooks, S. R., *Iron & Steel Eng*, **39** (1962), 72.
(12) Pfizer Trade Literature and Engineering (9 February 1968) 230.
(13) Sherritt Gordon Mines Ltd, Trade Literature.
(14) Fraser, R. W. and Evans, D. J. I., *Powder Met.*, **11** (1968), 330.
(15) International Nickel Technical Literature also Worn, D. K., and Perks, R. P., *Powder Met.*, **3**, 45.
(16) Stephens, D. M. and Greetham, D., *Powder Met.*, **11** (1968), 330.
(17) Michalke, M. and Scholz, W., "Proceedings of 2nd European Symposium on Powder Metallurgy Stuttgart 1968", Vol. 1.
(18) "Water Atomisation", A. O. Smith, Corporation Trade Technical Literature.
(19) "Gas Atomisation—Water Atomisation". Mannesmann Pulvermetall Trade Technical Literature.
(20) "Sponge Iron Powder," Hoganas AB Trade Technical Literature.
(21) "Powder by Gaseous Reduction," Pyron Co Trade Technical Literature.

(22) Gravenor, C. P., Govett, G. J. and Rigg, T., *Can. Min. Met. Bull.*, **624** (1964), 421; see also Canadian Patent 726,645 and Finlayson, P. C., *Metallurgia*, **77** (1968), 201.
(23) "Electrolytic Power," Cohen 600 Group "Sintrex" Technical Trade Literature.
(24) "Carbonyl Powders," International Nickel—Technical Trade Literature.
(25) Stevens Daugherty, T., *Powder Met.*, **11**, 342.
(26) Anon., *Iron and Steel Eng*, **39**, No. 2 (February 1962), 72.
(27) Metal Powder Association Standard, 3–45.
(28) Sturgeon, G. M., Jackson, G., Barker, V. and Sykes, G. M. H., *Powder Met.*, **11**, 314.
(29) US Patent 3,121,631.
(30) British Patents 1,006424 and 1,115723.
(31) Davies, I., Gibbon, W. M. and Harris, A. G., *Powder Met.*, **11** (1968), 295.
(32) Jaffrey, W. G., Davies, I. and Taylor, R. L. S., *Science Jnl*, **5A**, No. 3, 61.
(33) Delegate Contribution—International Powder Metallurgy Conference, New York (1970).
(34) Niesz, D. E., Fleck, J. N. and Kistler, Jr, C. W. AIAA Paper, 68–338, AIAA/ASME, 9th Structures, Structural Dynamics and Materials Conference, California (1968).
(35) Porembka, S. W., Fleck, J. N., Niesz, D. E. and Kistler Jr, C. W., AIAA Paper, 5th Aerospace Sciences Meeting, New York (1967).
(36) C. E. News Feature, "Marriage of Materials Gives New Steel Foil Products" *Chemical Engineering* (11 March 1968), 100.

INDEX

NEWCASTLE UPON TYNE
POLYTECHNIC LIBRARY

NEWCASTLE UPON TYNE
POLYTECHNIC LIBRARY